ATM and Internet Protocol

A convergence of technologies

Mark Bentall
Cogent Defence Systems
Newport, Gwent
UK

Chris Hobbs
Nortel Enterprise Networks
Ottawa
Canada

Brian Turton
Division of Electronic Engineering
Cardiff School of Engineering
University of Wales
Cardiff
UK

A member of the Hodder Headline Group

LONDON • SYDNEY • AUCKLAND
Copublished in North, Central and South America by
John Wiley & Sons Inc., New York • Toronto

First published in Great Britain in 1998 by
Arnold, a member of the Hodder Headline Group,
338 Euston Road, London NW1 3BH
http://www.arnoldpublishers.com

Copublished in North, Central and South America by
John Wiley & Sons Inc.,
605 Third Street,
New York, NY 10158-0012

Whilst the advice and information in this book is believed to be true and
accurate at the date of going to press, neither the author nor the publisher
can accept any legal responsibility or liability for any errors or omissions
that may be made.

British Library Cataloguing in Publication Data
A catalogue record for this book is available from the British Library

Library of Congress Cataloging-in-Publication Data
A catalog record for this book is available from the Library of Congress

ISBN 0 340 71921 4
ISBN 0 471 31427 7 (Wiley)

Commissioning Editor: Sian Jones
Production Editor: James Rabson
Production Controller: Rose James
Cover designer: Terry Griffiths

Typeset in 10/12 pt Times by Focal Image, Torquay
Printed and bound in Great Britain by J W Arrowsmith Ltd, Bristol

*This book is dedicated to the memory of
Isambard Kingdom Brunel,
an engineer*

CONTENTS

PREFACE

The world's telecommunications networks were created to carry voice traffic but are now being called upon to carry various types of data, typically between computers. 1996 is believed to have been the first year when data traffic exceeded voice traffic.

Over the last hundred years, many large telecommunications companies have been established to transfer voice from one customer to another. Such companies are seeing their traditional markets being eroded and are, in general, responding as quickly as their size allows. Broadband ISDN (B-ISDN) is a key element in their strategies.

There are also large companies which produce the devices required to transfer data. Now that data traffic predominates, and since voice can be considered nothing more than a particular kind of data, business that was traditionally the sole preserve of the telecommunications companies is now being targeted by their new competitors whose weapon is the Internet Protocol – IP.

As these two forces square up to each other, the interplay of B-ISDN and IP has become a crucial factor on the battlefield.

This book attempts to explain B-ISDN to a reader entering the modern telecommunications field either from the data arena or from the more traditional telecommunications disciplines (X.25 or Frame Relay). It then explores the interrelationship of IP and ATM. Little more than a basic understanding of telecommunications is assumed from the reader.

As with most areas of modern telecommunications, B-ISDN is complex, tightly coupled to other techniques and encompasses a forest of abbreviations, jargon and acronyms. The best advice that the authors can give a newcomer to this field is to have been born in about 1950. This would bring them into the telecommunications industry in the mid 1970s, a time when the installed systems were still relatively simple and the techniques which now dominate the field were still in their infancy.

Anyone statisfying this requirement would have been able to participate in the definition of the ISO OSI reference model in the late 1970s, produce packet switching devices to the X.25 standards and collaborate with the IEEE on the creation of the LAN standards in the early 1980s, work with the emerging digital voice switching standards in the mid 1980s, design a few ISDN terminal adaptors in the later 1980s and see the labyrinth of later standards – on Frame Relay, ATM and Network Access – emerge from the CCITT and ETSI in a reasonable historical sequence.

Readers born significantly later than 1950 are coming into an area of enormous complexity without this historical perspective and are likely to find themselves among experts talking apparent gibberish and quoting from thousands of standards documents, each hundreds of pages long and each referring in turn to several dozen others.

This is the state of the telecommunications industry today. A few years ago, B-ISDN and ATM, the primary subjects of this book, were envisaged as unifying and simplifying technologies, designed to remove the layers of baroque complexity which had grown in the previous 15 to 20 years. In practice, by combining speech, data and video on one network, B-ISDN has sparked the battle for survival between the large, slow-moving telephone companies, bent on deploying ATM and mopping up the data traffic business, and the small, fast-moving data companies, meeting the challenge by moving into the speech transmission business. The battle is not yet over and it is not yet obvious who will win.

This is an exciting time in the telecommunications field. The purpose of this book is to lead readers gently onto the battlefield and, by introducing the techniques of B-ISDN in relation to IP, to arm them for the fight.

This book includes a glossary, starting on page 129, with short tutorial articles on terms of general telecommunications applicability.

Mark Bentall
Cogent Defence Systems, Newport, Wales

Chris Hobbs
Nortel Enterprise Networks, Ottawa, Canada

Brian Turton
University of Wales, Cardiff, Wales

1
AN INTRODUCTION TO B-ISDN AND ATM

Readers with little telecommunications background may find the glossary entry for telecommunications *on page 154 useful when reading this chapter.*

1.1 Broadband ISDN

This chapter gives an outline of Broadband ISDN and ATM, an outline which is amplified in the following chapters.

The road towards ATM began in the 1980s with Packet-Switched Networks based on the X.25 standards (see reference [1]). In these networks, data traffic is *X.25* broken into variable-length frames and the frames are routed individually across the network for re-assembly at their destination. In the slow and unreliable networks for which X.25 was designed, a fairly heavy protocol (i.e. a protocol requiring substantial computing power and carrying a significant overhead with the data) was required to ensure that errors in transmission affecting one frame did not require whole files to be retransmitted.

Networks are now much more reliable – the bit error rate probability of a modern fibre optic link is approximately 10^{-7} to 10^{-12} per bit (see references [2, 3, 4]) – and with this improvement, it is possible to lighten the protocol and switch variable-length packets very quickly; the new, lighter protocol is known as Frame Relay.   *Frame Relay*

Frames, however, have different lengths and even when associated with a light protocol, they are difficult for hardware to handle; they generally require software intervention and this slows down their transit through switches. A logical step, therefore, was to fix the size of the frame (and to call it a *cell*), *cell* and ATM, the transport mechanism for Broadband ISDN, was born, combining a lightweight protocol with fixed-length cells. It should be said, however, that due to the inflexibility of fixed-length cells, there now appears to be some pressure on integrated circuit manufacturers to support variable-length frames in hardware and a resurgence of frame – rather than cell – switching is a possibility in some environments.

Broadband ISDN (B-ISDN) is a set of communications protocols designed *B-ISDN* to transport a wide range of services simultaneously, services which previously traversed different networks. In general, these services require more bandwidth than is offered by a traditional speech telephony circuit and may eventually include residential services such as distributed video (including home shopping, TV, broadcast distance learning, high-speed Internet connections, video-on-demand, catalogue and advertising services) and interactive multimedia (such as home doctor, virtual reality, interactive games, individualised distance learning and on-line yellow pages).

While the economics of residential B-ISDN still requires a great deal of study (for example, it is easy to calculate that, scaled against the cost of a conventional telephone call, watching a 2 hour video by this method should cost in the region of £1500), the business applications of B-ISDN make more immediate financial sense. Although most of the popular publicity has concerned broadband delivery to the home, while some companies are *trialling* broadband to the home, many more are *selling* B-ISDN to businesses.

Business applications include the interconnection of local area networks, multimedia telephony and teleconferencing, computer load-sharing, working at home (tele-commuting) and the transmission of images to centralised pools of experts.

Of course, these services have been available for some time, but only by using a plethora of different devices (routers, bridges, etc.) and transmission techniques (leased lines, public packet switching networks, etc.). The purpose of B-ISDN is to simplify this for the business customers while reducing their costs – instead of renting a permanent leased line between offices a company can now access the B-ISDN services and pay only for usage.

MAC

ATM

Underlying B-ISDN is the low-level Media Access Control Protocol (MAC) for transferring the actual information. This is common to all services and is known as Asynchronous Transfer Mode (ATM). B-ISDN and ATM are commonly used synonymously, but the distinction is quite clear: B-ISDN is the means of carrying traffic seen by the end-user, ATM is the low-level protocol which actually does the carrying.

Thus, several trends are encouraging the widespread introduction of ATM: the availability of high-speed, low error-rate communication links between switching centres, the availability of technology to digitise video and speech and the pressure to reduce operating costs by integrating previously separate telephony and data networks.

1.2 Traffic types

Speech has been digitised for transmission across the voice network for many years; intercomputer data traffic is, by its nature, digitised; and with the more sophisticated standards and technology for compressed video (and with consumers lowering their standards of video acceptability), video can now also be digitised within a reasonable bandwidth. Thus, three services which have traditionally been carried across different networks are now simply strings of numbers which need to be transmitted from point to point.

This is, of course, a very simplistic view; although the information for each of the services is a stream of numbers, the necessary characteristics of their transmission differ greatly:

- Speech services must have a reasonably short and constant delay between source and destination. Long delays, such as those sometimes encountered on trans-Atlantic satellite links, make two-way conversation awkward and large variations in delay produce broken sound effects which make speech difficult to understand. A generally accepted maximum delay is 25 ms and delays longer than this are not only unacceptable to users, they also make echo-suppression equipment very complex and expensive.

  It is also very important to preserve the order in which audio samples are received; if the samples are misordered in transit then they have to be discarded or buffered, causing delay and increased cost at the receiver.

  However, the situation is not all bad: humans are more tolerant of interference on signals than is a computer. The analogue telephone network reduces audio quality significantly in any case and subscribers are familiar with this. This tolerance of subscribers to reduction in speech quality is also exploited by carriers who reduce the bandwidth on their lines by compressing the speech and by not transmitting during silent periods. Since over 50% of a telephone conversation in each direction is normally silent, this technique, known as *Silence Suppression*, reduces the amount of information to be sent at the cost of turning what was a constant bit-rate datastream into a variable bit-rate stream.

- Video broadcast services have characteristics similar to speech though the rate at which information is transported is much higher and absolute delay is no longer of primary importance. Delay variance, however, and misordering of received packets is still a major problem although a limited amount of such distortion is acceptable to human consumers watching the result.

  To reduce the network bandwidth, compression techniques (see, for example, reference [5]) are used to remove repetition within and between successive video frames. Using compression produces a varying data rate; for example two successive similar video frames would require fewer data to be transmitted than two very dissimilar frames. Within the telecommunications community, the video of the film *Star Wars* has become almost a *de facto* standard for test purposes because of its sudden (and violent) changes in colour and background. The use of compression with video, as with audio, has the characteristic of turning what is essentially a constant bit-rate signal into a variable bit-rate stream. Data rates of between 1.5 Mb/s and 8 Mb/s are typical.

- Inter-computer traffic includes document text, data files, still pictures, audio and video. Generally, this type of transmission is sensitive neither

Service	Absolute delay	Delay variance	Corruption	Misordering
Audio	Must be small	Intolerant	Tolerant	Intolerant
Video	Tolerant	Intolerant	Tolerant	Intolerant
Inter-computer	Tolerant	Tolerant	Intolerant	Tolerant

Table 1.1 *Service characteristics*

to delay nor to delay variance since nothing is done with the data until they have all arrived and the original file has been re-constructed. The receiving computer is usually also tolerant of mis-ordering of the parts of the message in transmission, being able to hold parts which arrive early and insert them into their correct position later. The computers are generally, however, intolerant of any corruption of the information being transferred: these services rely on an error-free connection.

All services can be considered as a combination of these three elements: real-time audio, real-time video and inter-computer data. The characteristics of these basic elements are summarised in Table 1.1 where it can be seen that providing the necessary Quality of Service to each of these simultaneously is likely to prove difficult.

One characteristic of these services, hinted at above but not included in this table, is the variability of the bit rate. Four different bit rate types are supported by the ATM standards: Constant Bit Rate (CBR), Variable Bit Rate (VBR), Available Bit Rate (ABR), and Unspecified Bit Rate (UBR). These are described

CBR

in more detail below but, as the name suggest, CBR traffic derives from a source where information is presented at a constant rate (for example, telephony speech without silence suppression).

VBR

VBR traffic derives from a variable source (for example, compressed video or voice with silence suppression). Two types of VBR have been defined:

VBRrt
VBRnrt

VBRrt (VBR real time) and VBRnrt (VBR non-real time) to handle traffic with the respective characteristics. A great deal of research has been carried out in recent years into the observed very 'bursty' nature of some data traffic. Much inter-LAN and compressed video traffic has been shown to be *self-similar* or *fractal*: that is, over whatever time interval the traffic is observed, the same type of 'burstiness' appears. Thus the bursts themselves appear in bursts. This type of traffic causes significant difficulties for practical switches because of the difficulty in estimating (and supplying) the correct amount of buffering. Since the beginning of telephony, assumptions have been made about the aggregation of traffic: for example, assuming that while the traffic from one telephone source is likely to be somewhat unpredictable, the combined traffic from thousands of telephones is highly predictable and stable. This, so-called, Poisson model breaks down very badly with the self-similar traffic where there is little or no 'averaging' of the peaks and troughs from multiple sources. See reference [6] for more details of this very important phenomenon which was observed and recorded as early as 1951, albeit not in ATM networks (see reference [7]).

ABR is a later addition to the ATM world. Once a carrier has allocated the *ABR* necessary bandwidth on his links to carry the CBR traffic and the guaranteed minimum for the VBR traffic, ABR is the mechanism to share the remaining bandwidth fairly, between subscribers. To support ABR, a customer's equipment must accept signals from the network to indicate that it should reduce or increase the bandwidth it is using (that is, send cells more slowly or more quickly). This type of traffic rate, while convenient for an application where the transfer of data is neither delay- nor time-critical, places high demands on the network but allows the network providers to operate the network highly efficiently. ABR has also been termed Best Effort Delivery and is expected by some to replace the Internet *Best Effort* Protocol (IP) since it offers higher transmission speeds (see reference [4]). The *Delivery* mechanisms whereby the network signals congest to the attached terminals have been another area of hot debate – between the rate-based schemes, which would appear to work better in wide area networks, and the credit-based schemes, which work better in local area networks. The basic principles of the two schemes are:

- In a rate-based scheme, the network provider signals a maximum bit-rate *Rate-based* to the user's terminal and the terminal is not allowed to transmit at a *Flow Control* higher rate. When the network congests, it signals a lower rate to each of the sources; as the congestion clears it signals a higher rate.
- In a credit-based system, the terminals and switches in the network ex- *Credit-based* change so-called *credits* with their neighbours and only devices in posses- *Flow Control* sion of credits may transmit. If the number of free buffers in a particular switch is becoming dangerously low, that switch will withhold credits from its neighbours. The neighbours will then not be able to send it ABR traffic.

For a time it appeared that a compromise between these would prevail, using rate-based algorithms across the wide area and credit-based algorithms in local areas, but this was finally abandoned in favour of rate-based techniques. This scheme relies on Resource Management (RM) cells which are generated *RM* artificially by the source device and which are inserted into the data stream. These cells are reflected by the destination back onto the originator. As the RM cells return, they are modified by intermediate switches which are experiencing congestion to indicate to the source that it should reduce its flow rate.

UBR is the simplest technique of all to describe: there are no guarantees at *UBR* all for this sort of traffic. The bandwidth provided is left to the discretion of the network provider who makes no guarantee about the bandwidth, traffic delay or the likelihood of traffic loss. If there is any flow-control in UBR then it must come from the end-devices (by using, for example, a protocol such as TCP) because it is not provided by the network. This mode of working has become a *de facto* standard as it is the only one that many ATM switches originally offered.

The various classes of service can be compared (loosely) with the different ways one can purchase theatre tickets:

- CBR is equivalent to booking in advance a given number of seats for every performance: the seats are prebooked and their availability is guaranteed
- VBR is similar to arranging with the theatre that it will make available an average of a specific number of seats for every performance, but also specifying a maximum number of seats which will be required
- ABR is equivalent to arranging to take a certain minimum number of tickets for each performance but with the agreement that, if more seats become free for a particular performance, then they will be made available
- UBR is equivalent to turning up without tickets to each performance and trying to get in.

Services can be monomedia or multimedia depending on the information content. For multimedia services there are additional requirements to be met concerning synchronisation between the media (e.g. between the sound and the picture of a film).

1.3   ATM

The ATM protocol itself is not concerned with the content of the frames it transports. It breaks the frames into fixed-length cells which it transports across the network, the frames then being reconstructed at the destination. The protocols which break the frames into cells are known as the ATM Adaptation Layers (AALs) since they adapt frames from different media to and from the ATM network.

AAL

To allow one network to be shared between many users, the ATM protocol sends a small amount of housekeeping information across the network with each cell (containing, for example, the identity of the path that the cell is following and the type of data it contains).

By converting traffic into fixed-length cells, ATM allows users to transmit any combination of video, voice or data traffic without disrupting other users' services. ATM being eponymously asynchronous, it does not allocate a specific time between the transmissions of successive cells or frames; instead it serves each user to the best of the network's ability.

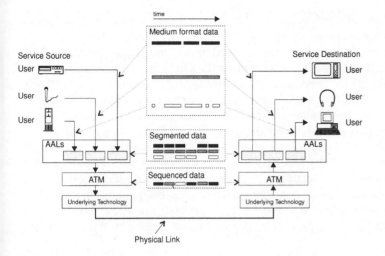

Fig. 1.1 *Basic data flow*

Figure 1.1 shows an example of how different types of traffic co-exist on the same network – each user's information is broken into cells before transportation, irrespective of the service being carried. Note that, for clarity, this figure shows unidirectional traffic whereas, in reality, useful traffic is almost always bidirectional. The figure illustrates:

- The different natures of computer-to-computer traffic, video traffic and telephony (speech) traffic.

  Video traffic uses compression algorithms which mean that each video frame varies in size. While a delay of even a second can be tolerated in the transmission of the video, delay variations cannot be tolerated as they would lead to the break-up of the picture.

  Uncompressed speech produces information continuously and at a constant rate – both absolute delay and delay variance are important.

  Computer-to-computer data traffic (for example a file transfer) produces variable-length frames at irregular intervals. Neither absolute delay nor delay variance is important.

- The concept that the user's traffic passes through an ATM Adaptation Layer (AAL) where it is broken into ATM cells and from this point onwards, until the destination AAL reconstitutes it, no attempt is made to keep the cells from the individual users separate. The ATM function at the receiving end uses the information sent with each cell to separate out the three traffic paths and the AAL there reconstitutes the original frames.

Once it was decided that the ATM cell size should be universally fixed, the length had to be chosen and agreed by all parties. This was not easy and the final decision is imperfect. To transfer delay-sensitive services such as speech, it is important that cells be reasonably short. They would otherwise take too long to fill and the network operator would have to transfer uneconomical, half-empty cells across the network or wait for the cell to fill, making the end-to-end delay unacceptable. Delay-insensitive traffic such as computer-to-computer file transfers, on the other hand, works best with long cells. With short cells the overhead of the housekeeping information sent with each cell is high.

Given these two pressures (and a great deal of trans-Atlantic, in particular France/USA, rivalry), the choice of cell size was bound to be a compromise and at a meeting of the ITU in Geneva in 1988, two options were tabled: 32 octets (bytes) and 64 octets. Finally a data length of 48 octets was chosen with a 5 octet header for housekeeping (this, approximately 10%, being the greatest overhead considered acceptable). It is instructive to calculate the delay introduced into speech traffic and the overhead introduced into data traffic by this compromise choice:

- Normal telephony-quality speech is converted into one octet every 125 μs. Thus $48 \times 125 = 6000$ μs are required to fill the 48 data octets of an ATM cell, thus introducing a 6 ms delay to each sample in addition to network switching and transmission delays. A delay of 6 ms is just acceptable but, when the speech is compressed to 32 kb/s or even 16 kb/s to reduce transmission bandwidth, the time taken to fill a cell increases to 12 or 24 ms. This, when combined with switch delays, being unacceptable, either cells must be transmitted part-filled (thus defeating the objective of reducing bandwidth by compression) or samples from multiple speech channels must be carried in the same cell. The latter technique introduces complexity into the switching since only speech channels with a common source and destination switch can be so combined.

- Data traffic is broken into, at best, cells of 48 octets each with a 5-octet overhead. This means that at least $5/53 = 9.4\%$ of the link capacity is not available for useful data transfer.

Cell Tax

The $5/48 = 10.4\%$ overhead added to each cell has been called the 'Cell Tax': the price one pays for using ATM.

Channel

Cells carrying speech and video must be received in the order they were sent. This is known as preserving *data integrity* and is a function of the ATM layer. Any link which preserves the order of data entering and leaving is known as a *channel*. To achieve this within B-ISDN, a route (known as a *Virtual Channel Connection*) is established through the network and all the cells follow this same route, thereby preserving cell order. A protocol that offers this type of connection between source and destination is known as *connection-oriented*.

Connection-Oriented and Connection-less Protocols

There are actually two concepts intermixed and often confused in the idea of Connection-Oriented and Connectionless Protocols: path and pre-established connection. There are protocols (for example the User Datagram Protocol (UDP)) where no end-to-end connection is ever established and the traffic follows no definite path. Each packet of information is simply launched into the network to find its own way to the possibly nonexistent destination. There are protocols (such as TCP) where an end-to-end connection has to be established before traffic can start to flow but for which there is no definite path through the network. Finally there are ATM-like protocols where an end-to-end connection is established before traffic can start to flow and all traffic then follows the same path through the network (unless switches or links fail, causing the connections to be moved to non-failed equipment). It is only when a path is predefined and used by all traffic that true Quality of Service guarantees can be given to the application using the connection.

Since ATM is inherently connection-oriented, when connectionless protocols need to be carried across the ATM network, they are implemented by an ATM Adaptation Layer (AAL).

1.4   Virtual Connections

In a connection-oriented network a *Virtual Channel Connection* (VCC) (some-   *VCC* times loosely called a *Virtual Circuit*) is created between source and destination before traffic can flow.

The connection is virtual since it consists of many point-to-point links between switches and does not physically exist as a whole. To the users at each end of the virtual channel connection, however, it has all of the necessary properties: information poured into one end appears, in order, at the other end. A particular Quality of Service is associated with a VCC when it is established.

A VCC exists from the traffic source to its destination but it actually comprises a number of concatenated Virtual Channels, each spanning between two switches.

To allow Virtual Channels to be identified, each is allocated a number, or identifier, unique within a switch.

When several VCCs are established between a pair of users it is convenient to group them into a single Virtual Path Connection (VPC). Similarly Virtual   *VPC* Channels are grouped into Virtual Paths (VPs). Figure 1.2 (taken from I.113 –   *VP* reference [9]) illustrates the concept of Virtual Channels within a Virtual Path and multiple Virtual Paths being carried over the same transmission link.

Note again that Virtual Channels and Virtual Paths have no physical reality: the cells for all channels and paths are sent, without regard to order, along the same transmission link and it is only the Virtual Channel Identifier and Virtual Path Identifier carried in a cell's header that allows it to be identified at the destination.

One disadvantage of a connection-oriented network is the amount of state information which must be held within the switches of the network. Details about each Virtual Channel Connection will be held in every switch through which it passes. This ultimately leads to a scaling problem when many connections are

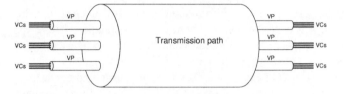

Fig. 1.2 *Virtual Channels and Virtual Paths*

established, partly because of the amount of stored state information and partly because of the time taken to re-establish all of the connections following the failure of a node. Connectionless protocols do not suffer from this explosion of information but instead require time-consuming routing to be carried out on each packet at each switch. Section 6.2 contains details of approaches that aim to combine the speed of the connection-oriented switching with the flexibility of connectionless routing.

Signalling   Protocols, known as signalling protocols, are used to establish virtual connec-
Protocol   tions and to assign the identifiers. Once a virtual connection has been estab-
lished, intermediate switches receiving a cell use the address in the cell header to switch the cell to the appropriate physical (i.e. real) output port.

1.5   The ATM Switch

In the general descriptions given above, the concept of an *ATM Switch* has been used but not explained. Physically an ATM Switch consists of one or more shelves of equipment, typically with plug-in cards providing computing resources to control the shelf, interfaces to the various links and the actual
Fabric   switching fabric. An example of a small frame/cell switch, the Passport 6480 from Nortel, is shown in Figure 1.3.

The switch is designed to be fitted into a standard-sized rack and consists, from bottom to top, of a cooling drawer forcing air upwards through the entire switch to keep the electronics cool; three power supplies, any two of which can power the entire switch; 16 cards and a cable management area. Notice the quarter-circle fibre management mouldings at the top of the shelf – these are to prevent the incoming fibres being bent to too small a radius, causing damage to the fibre. The 16 cards include two shelf processors (one acting as standby for the other) which control the shelf, detect and report faults, offer an interface for switch management and hold the configuration of the shelf. The remaining 14 cards are interface cards chosen to satisfy the customer's demands: possibly LAN cards and a single WAN card in a small private network or a collection of Frame Relay and ATM WAN cards in a large carrier network.

Fundamentally, an ATM Switch such as that shown in Figure 1.3 accepts incoming ATM cells from its input links and switches these cells to the appro-

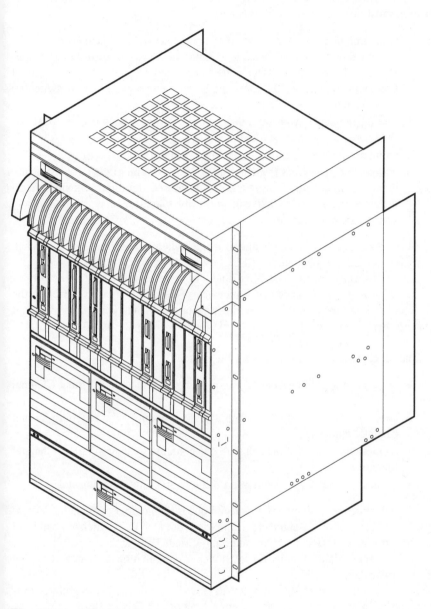

Fig. 1.3 *An ATM Switch*

priate outgoing links, taking into account the relative priorities of the cells from different Virtual Channels. This process is more complex than it sounds as the switch must also be prepared to receive:

- signalling messages from users requesting that Virtual Channels be created or torn down. One measure to compare switches from different manu-facturers is the rate at which they can establish and tear-down Virtual Channels: a rate of 100 connections per second per Gb/s of switching capacity represents the current state of the art
- routing messages from the other switches in the network indicating that links have failed or recovered
- maintenance messages from network operators
- maintenance messages from its constituent electronics indicating that equip-ment has failed, requiring traffic to be passed through the switch along different paths – a switch will generally tolerate and work around any single hardware failure.

These tasks are further compounded if the switch is not a pure ATM Switch – if, for example, it also handles X.25, Frame Relay or IP traffic.

In order to carry out its switching function, a switch must have some way of knowing how to handle each cell it receives. This information is carried in the 5-octet header of each cell, in a field identifying the Virtual Channel on which it has arrived. The switch has a table listing the outgoing port and Virtual Channel for every incoming Virtual Channel.

Therefore, when a cell is received, the switch:

- examines the cell's header to discover to which incoming Virtual Channel it belongs
- looks up that Virtual Channel in a table to identify the outgoing Virtual Channel and Port
- replaces the header of the cell with one containing the new Virtual Channel Identifier
- queues the cell for transmission on the appropriate outgoing link.

The fundamental limitation of this technique is the number of available Virtual Channel Identifiers. To avoid this limitation, some ATM Switches switch all traffic on one Virtual Path to an outgoing Virtual Path without examining the Virtual Channel field. Figure 1.4 illustrates a combination of Virtual Channel and Virtual Path Switching.

ATM Switches are highly complex devices but their hardware architectures tend to be of one of two types: Common Memory or Output Buffered.

Both architectures include line interface cards of various types, a control element and some means of passing cells between the interface cards. In practice switches need to be highly reliable and, generally, all hardware elements of the switch will be duplicated to meet unavailability times of under 1 minute per

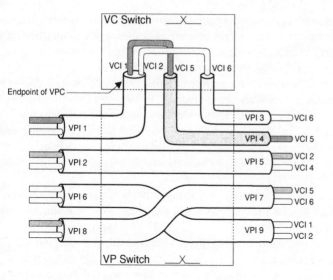

Fig. 1.4 *Virtual Channel and Virtual Path Switching*

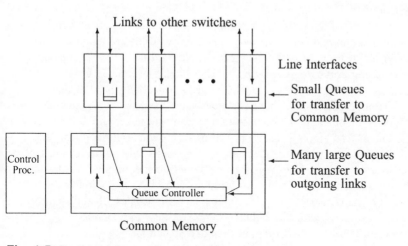

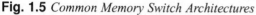

Fig. 1.5 *Common Memory Switch Architectures*

year. Hardware failure at the switch is only one of the possible causes of switch unavailability: software errors, scheduled outages for software upgrades to be installed and failures of outside plant (cables and fibres between sites) also contribute to this. Manufacturers are now producing switches which do not need to be taken out of service to install new software and, apart from cables being dug-up and broken, the major causes of switch failures are now software bugs. With switches containing an increasing amount of software, the proportion of these failures is likely to increase and the classification and prediction of software failures is much less understood than the hardware equivalents.

In a small switch which converts between frames and cells at the edge of the network, a customer's devices are connected to low-speed cards offering interfaces such as E1 and E3 (refer to the glossary entry for *plesiochronous*), possibly channelised to DS-0. Such a switch would probably be capable of switching cells at rates of about 5 to 10 Gb/s and would use one or more STM-1 links (again see *plesiochronous* in the glossary) towards the core of the network.

In a larger, pure cell-switch in the core of the network, interfaces are likely to be at, OC-12, OC-48 and OC-192 rates and switching speeds in the region of 160 Gb/s to 1 Tb/s.

Once the actual transmission lines have been terminated, the received cells have to be stored somewhere to await their turn at the output link (several input links could be connected to the same output link and all inputs could be supplying bursts of traffic at the same time). Additionally, if the switch is an edge switch then received cells may have to be stored until a complete frame can be reassembled. It is the location of this storage which differentiates the Common Memory and the Output Buffered architectures.

A Common Memory Switch as shown in Figure 1.5 has a single, central memory for all of the cells it is temporarily storing. The interface cards simply transport cells to and from the common memory. Cells are typically 'chained' together in queues in the common memory, older switches having one queue for each outgoing link but more recent switches having one queue for each Quality of Service on each outgoing Virtual Channel – so-called per-VC Queuing.

Per-VC
Queuing

An Output Buffered Switch, on the other hand, has only a very small common memory (sometimes no memory at all and at most memory for a few hundred cells) but instead stores cells on the card which will transmit them (see Figure 1.6). The incoming interface card will store cells only until it can pass them to the outgoing card. They will then be stored on the outgoing card, possibly with one queue for each Quality of Service for each Virtual Circuit.

As with many aspects of telecommunications, debates of almost religious intensity take place about the relative merits of these two architectures.

The main merits of the Common Memory Switch are:

- that the line interface cards are very simple (and therefore cheap) since they require little storage and no traffic-handling intelligence
- that this type of switch is probably easier to design
- that multicasting (the sending of the same cell to multiple destinations) is simpler as the cell simply needs to be put logically onto several outgoing queues
- that hardware redundancy (to protect against failure) is easier to build in.

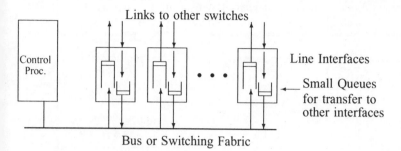

Links to other switches

Control Proc.

Line Interfaces

Small Queues for transfer to other interfaces

Bus or Switching Fabric

Fig. 1.6 *Output Buffered Switch Architectures*

The main merits of the Output Buffered Switch are its flexibility (the amount and speed of cell memory can be tuned to the particular output ports) and its low entry cost. A Common Memory Switch requires a large number of extremely fast – and expensive – memory chips even in its smallest configuration; the Output Buffered Switch may actually have more cell memory because it is fragmented but the speed and size of the memory can be more easily tuned.

In recent years there has been a swing away from Common Memory Switches towards Output Buffered Switches.

1.6 Service Classes

The variability of the bit rate and the connection mode (connection-oriented or connectionless) are important characteristics of an end-to-end path. Another important factor is whether timing information needs to be preserved from source to destination (i.e. whether information delivered at the destination must appear at the same time intervals as when it was presented by the source). These three important characteristics together define a *Service Class*. An example of *Service* a service requiring the timing relationship to be transferred across the network *Class* would be that of a customer currently using a leased line (a rented real line) to transfer data synchronously between two devices (using, for example, the HDLC protocol). In order to reduce costs the customer may want to replace the leased line by a connection across an ATM network but, to avoid having to buy new terminal equipment, the synchronous nature of the traffic must be preserved.

Since not every combination of the three basic characteristics is meaningful, the ITU originally defined four composite Service Classes (see I.362 – reference [10]) known as classes A to D as shown in Table 1.2. Since each class of service requires an ATM Adaptation Layer (AAL) to convert between the user's traffic format and ATM cells, these four classes were originally mapped to AAL-1, AAL-2, AAL-3 and AAL-4.   *AALs*

	Class A	Class B	Class C	Class D
Timing synch. source/dest.	Required		Not required	
Bit rate	Constant	Variable		
Connection mode	Connection-oriented			Connectionless

Table 1.2 *Service Classes*

Understanding the evolution of the Service Classes and the numbering of the AALs requires enormous patience. For example, it soon became clear that, from the point of view of the AALs, classes C and D were identical.

Thus three AALs were defined: AAL-1 for class A, AAL-2 for class B and AAL-3/4 for classes C and D.

More recently AAL-5 (now the fourth AAL) has been defined, officially to cover class C but actually to cover classes C and D. Whether class D traffic can use AAL-5 is a matter of personal belief rather than a technical debate. Strictly, since class D is connectionless and all ATM connections are connection-oriented, class D cannot be supported by ATM. In practice, however, the characteristic of connectionlessness is always supported at protocol levels higher than ATM (normally by opening several connection-oriented pipes) so this becomes a non-issue.

AAL-CU
A new adaptation protocol, commonly known in Europe as AAL-6 and elsewhere as AAL-CU, was suggested within the ATM Forum (see glossary entry on page 131) to allow multiple subcells to be carried within one ATM cell. This protocol, which breaks the basic rule of ATM that all cells be of fixed length, was expedient for carrying low-speed data where the delay caused by waiting for a cell to fill was too long and the overhead of carrying an incomplete cell was too great. Eventually, finding that AAL-2 had not been defined, the ATM Forum decided to call it AAL-2. This has very little to do with the original intention for AAL-2.

More detail is given for each of these ATM Adaptation Layer interfaces in Chapter 3 starting on page 39.

Since error performance is service-specific (detection of bit errors, for example, being more important to computer-to-computer data exchange than to a speech connection), it is a function performed by the AALs. Errors affecting control information, for example addresses, are a concern of the ATM protocol itself.

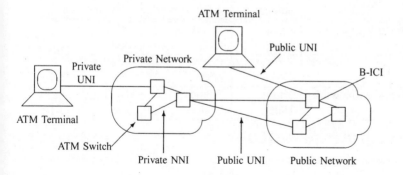

Fig. 1.7 *Network topologies*

1.7 Applications

The ATM protocol has been used for a wide variety of network topologies, but this book concerns itself primarily with telecommunications networks, that is networks having long distances between switches.

Telecommunications networks can be thought of either as carrier networks, run by national PTTs or private companies for public use, or as Enterprise *Enterprise* Networks, run by private companies or organisations for their own internal use. *Network*

With the fairly recent deregulation of the telecommunications market, particularly in Europe and North America, this previously useful division is becoming fuzzy, since the spare bandwidth on networks installed ostensibly for use within a company is often being offered for sale on the open market. The supply of long-distance lines between centres will probably remain with the carriers and a few specialised companies owning strips of land between cities: the railway operators, power companies, canal owners, etc.

ATM is likely to have most impact on Carrier Networks where the flexibility of being able to carry any sort of traffic is paramount. Within an Enterprise Network, where the owner has much more control over the data being carried, protocols such as TCP/IP arguably offer better throughput using cheaply available devices (routers, bridges, etc.) at less cost than ATM. This is particularly true for campus networks where links are typically short and topology is stable.

The use of TCP/IP in campus networks will require carriers to transport TCP/IP frames economically across an ATM network between sites. This topic is the subject of section 6.2.

Between	Terminal equipment	Private ATM network	Public ATM network
Terminal equipment	Private UNI	Private UNI	Public UNI
Private ATM network	Private UNI	Private NNI	Public UNI
Public ATM network	Public UNI	Public UNI	B-ICI

Table 1.3 *ATM reference interfaces*

1.8 Interfaces

At different points in the network, different types of ATM interface are defined as illustrated in Figure 1.7 and Table 1.3 (adapted from reference [11]):

- The interface between a user's terminal and the network. This type of interface is known as a User–Network Interface (UNI). Two subtypes of the UNI are shown in the diagram: between each user's terminal and his or her company's private ATM Switch, a *Private UNI* exists. Between the public ATM network and a terminal or private ATM Switch, the carrier offers a *Public UNI*.

UNI

- The interface between the carrier's network and the networks of other carriers, possibly in other countries. This type of interface is known as the *Broadband Inter-Carrier Interface* (B-ICI).

B-ICI

- The interface between switches which form part of a single larger private network is known as a *Private Network–Node* (or, sometimes, *Private Network–Network*) *Interface* (PNNI or NNI). The term PNNI is also sometimes used to refer to the routing protocol which occurs across a PNNI, the original version of which is known as the Interim (or Integrated) Inter-Switch Signalling Protocol (IISP). Not having to handle the complexities of splitting the customer's data into cells, this interface is somewhat simpler than the UNI and AAL protocols are not required.

NNI
PNNI

IISP

Note that Figure 1.7 does not specify the types of interface between the switches within a carrier's network. These interfaces are assumed to be private to the carrier and are not standardised.

To be connected to an ATM network, equipment must support the necessary UNI. The point in the customer's premises which defines the boundary (demarcation point) between the customer's equipment and the network is known as a Network Terminator (NT) and two types of network terminator, B-NT1 and B-NT2, are defined as shown in Figure 1.8. The interfaces shown in that figure are defined very precisely to ensure that equipment from different manufacturers will be compatible. In practice, they may be internal interfaces within a

NT

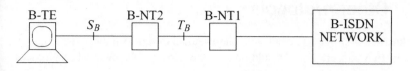

Fig. 1.8 *UNI functional architecture*

single piece of equipment. The definition of the precise interface between the customer's own equipment and the equipment owned by the network operator is very important. If a customer reports a fault then the network operator needs to be able to pinpoint very quickly whether it lies within its network or within the user's own equipment. To facilitate this, the B-NT normally contains a 'loop' which may be set by command from a remote site effectively isolating the customer's equipment, allowing the network operator to test the circuit as far as the B-NT.

The standards give several configurations for the B-ISDN UNI allowing it to be adapted to the bandwidth of customer and application requirements, for example a local area network (LAN) interface.

The primary functions of the B-NT1 are:

- physical termination of the transmission line (impedance matching, etc.)
- handling of the transmission interface (signal recovery, synchronisation, etc.)
- Operation and Maintenance (OAM) functions: (setting loops on command of the network operator to allow faults to be isolated, etc.).

The primary functions of the B-NT2 are:

- adaptation functions for different interface media and topologies
- multiplexing and demultiplexing if multiple user terminals are attached to the S_B interface
- buffering (at the ATM level). In practice, buffering is kept to a minimum in these devices (in some it is as little as three cells) because of the cost implications in devices intended for the end-user's premises
- signalling protocol handling
- OAM functions.

Note again that these two devices (B-NT1 and B-NT2) are artificial, created only to define their functionality; in practice they are normally combined into one real product.

1.9 Other protocols

Our discussion so far has dealt with the protocols used to transfer information between two subscribers. In addition to these protocols, two other types of protocol are needed for the network to operate:

- Signalling Protocols. A switching network based on a connection-oriented architecture must have protocols for determining the route and setting up a connection between the source and destination. The protocols that achieve this are known as signalling protocols and are discussed in Chapter 4.

- Control Protocols. Network operators must be able to control their network and gather sufficient accurate information to diagnose problems and to bill their users. The protocols which monitor equipment for failures and gather statistics and usage reports are known as OAM (variously rendered as *Operation, Administration and Maintenance* or *Operation and Maintenance*) (see I.610 – reference [12] and the whole of Chapter 5). OAM performs the 'FCAPS' functions for the network operator: Fault, Configuration, Accounting, Performance and Security Management. One weakness of the current B-ISDN and ATM standards is that they do not adequately cover the localisation of faults in a network. Each user of the network also expects (and pays for) a certain Quality of Service (QoS). Monitoring protocols ensure that this QoS is being maintained.

OAM

FCAPS

QoS

1.10 B-ISDN standardisation

ITU-TSS
ITU-T

ATM Forum

The International Telecommunication Union – Telecommunication Standardisation Sector (ITU-TSS or ITU-T) is currently standardising B-ISDN. Equipment manufacturers felt that the cumbersome conventional standardisation process would not allow a rapid introduction of technology into the market place and, to speed it up, formed a body called the ATM Forum, incorporating the commercial and academic establishments which are leading ATM development. Initially the ATM Forum published standards ahead of the ITU-T, allowing deployment of ATM technology in commercial products. Standardisation of the basic protocol structure of ATM is now complete and products meeting the ATM Forum standards are deployed. Now there is some evidence that the manufacturers are using the ATM Forum to slow down the introduction of new standards to allow existing products to be sold.

Having two standardisation bodies has, of course, led to problems, particularly between North America, where the ATM Forum standards are commonly used, and Europe, where customers want products meeting the ITU standards. For manufacturers trying to produce competitive products for sale in both markets, the differences are small but significant, particularly in the area of signalling protocols.

◇   ◇   ◇

In summary, the advent of Broadband ISDN and, in particular, ATM Switching, promises a significant simplification of the world's telecommunications networks: entirely separate networks carrying different services will be merged into one, high-speed, consolidated network. The underlying transport mechanism is also superficially simple: these complex services are converted into streams of constant-length cells at the edge of the network and switches are only required to forward these cells to their destination for reassembly.

This simplification in transport is, however, more than paid for in the complexity of handling cells that originate from services with different characteristics competing for bandwidth and buffer space. Many problems associated with the fair sharing of resources are yet to be solved.

When the speech, data and video networks were distinct the failure of one did not affect the others. With a combined network, a failure could, in principle, paralyse the telecommunications of a whole country. Manufacturers naturally build their telecommunications equipment to be resilient (duplicated processors, non-interruptible power supplies, etc.) and carriers build their networks to be redundant (installing duplicate links between key switches, etc.) but software errors can propagate themselves quickly through the most resilient network, bringing switches down like dominoes, and installation technicians can, and do, route the main and backup fibres through the same conduit, ready for both to fail simultaneously when the conduit is accidentally dug up.

The ways in which software errors manifest themselves are still not understood and it would be rash of anyone to say that an entire national telecommunications network could not be brought down by a single rogue line of software in a single ATM switch.

2
B-ISDN STRUCTURE AND ARCHITECTURE

2.1 The Functional Reference Model

The idea of a *Reference Model* for an architecture became popular during the *Reference* late 1970s when the International Standards Organisation (ISO) defined the Open *Model* Systems Interconnection (OSI) architecture by this method. Readers unfamiliar *ISO* with the layered architecture of the ISO OSI may wish to refer to the glossary *OSI* description of the term 'protocol' before reading this chapter. OSI standards were first published in 1984 under ISO 7490 and ITU-T X.200. There are many reference books on the subject, including [13].

OSI grew out of the proliferation of proprietary methods for connecting computers and tried to simplify the confusion by defining a number of functional layers, each performing a clearly defined task. The purpose of the standard was to define how the layers interact and the function (but not the implementation) of each layer. The complete set of layers is known as a *stack*.   *stack*

The highest layer, known as the *Application Layer*, contains the user's actual programs. The user is really only concerned that the application layer on one computer communicates with the application layer on another, possibly via several intermediate computers. In order to communicate, the software in the application layer makes use of the layer below it, the *Presentation Layer*. The Presentation Layer has no knowledge of the application that it supports but makes use of the layer below it, the *Session Layer*, to establish connection with the remote machine and to pass information back and forth.

The responsibility of the Presentation Layer is to ensure that the applications can communicate. The Session Layer knows what the remote application is. The first signs of networking do not occur until the *Transport Layer* where the remote application's location is determined. The lowest three layers are the key networking protocols: the *Network Layer* protocols determine the next step towards the remote application, the *Link Layer* protocol makes the next step towards the remote application and the *Physical Layer* knows how to use the physical medium for that step.

One principle is common to every layer (see Figure 2.1): each layer provides services to the layer above and relies on the services of the layer beneath, having

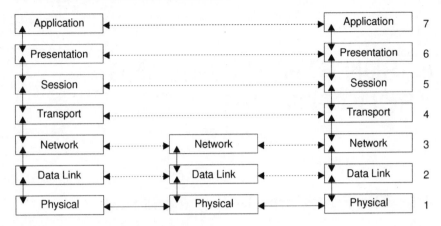

Fig. 2.1 *The ISO OSI Reference Model*

no knowledge of any other layers. Using the services of the layer below, each layer establishes a peer-to-peer dialogue with its equivalent layer in a remote machine, either at the destination or at an intermediate node.

Although the OSI, as originally defined, had a number of weaknesses (layers have since had to be subdivided and Network Management and Control have had to be added), the concept of a reference architecture onto which actual implementations could be hung is very powerful. The lack of Network Management in the reference model was a particular problem as it was quickly found that an effective management system requires access to all layers, thus contradicting the basic idea of self-contained layers communicating only with their adjacent layers by means of a small set of primitives.

B-ISDN has also been defined by means of a layered architecture (see Figure 2.2). Learning from the OSI experience, the horizontal layers have been divided into vertical planes. A great deal of debate has gone into the mapping of B-ISDN onto the OSI layers. This is very difficult and of limited usefulness once complete.

When different protocols interact, the corresponding stacks effectively sit alongside or on top of one another, making this type of diagram very difficult to interpret: a service with an ISO stack, for example, might make use of an ATM network for transport. In this case the whole of the B-ISDN stack, including the ATM Adaptation Layer, could form layer 2 of the original service. An example of interworking is shown in Figure 6.16 where a Frame Relay connection between two Frame Relay terminals makes use of a B-ISDN network for transport. At the appropriate points the information has to climb one stack and then descend the stack of the other service.

The vertical planes in Figure 2.2 represent the different functions occurring at each layer and are an addition to the ISO OSI model. All layers have a

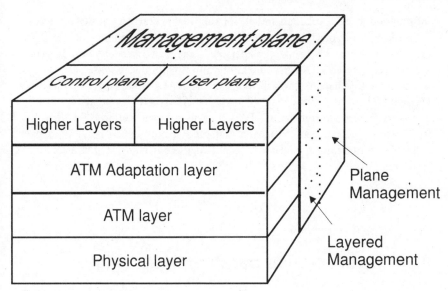

Fig. 2.2 *The B-ISDN Reference Model*

separate management plane which is further divided into plane management and layered management. Plane management incorporates the management functions relating to the whole system. Layered management relates to the management functions of a specific layer, and for the physical and ATM layers an OAM information flow is assigned; see Chapter 5. The higher layers have been further divided into a user plane for user information, and a control plane for call control signalling.

Starting at the top of the B-ISDN protocol stack the *ATM Adaptation Layer* *ATM* (AAL) more accurately corresponds to layer 4 of the OSI model and structures *Adaptation* the information in different ways in accordance with information from the higher *Layer* layers, depending on the type of service being carried. Many different service styles can be implemented by the AAL simultaneously. The AAL primarily segments the frames in the incoming data stream into ATM cells and re-assembles the cells into frames at the destination.

The *ATM Layer* provides routing information for the data passed to it from the *ATM Layer* ATM Adaptation Layer. Network management, signalling and the underlying OAM protocols are further functions of the ATM layer. ATM is an end-to-end protocol and relates most closely to the OSI layers 2 and 3, although OSI layers 2 and 3 are not end-to-end.

As with the OSI model, the *Physical Layer* is the point-to-point transfer mech- *Physical* anism sitting on top of the real hardware and may be as simple as a piece of *Layer* wire. The information from the ATM layer is received by the physical layer and transmitted to the next network node. The physical layer also adds its own information to each cell transmitted for link management purposes.

The lowest three ATM layers are defined by ITU-T recommendations. Sections 2.4, 2.5 and 2.6 below consider the ATM, AAL and Physical Layers respectively, from an ATM viewpoint. For an in-depth discussion of the different AALs refer to Chapter 3 which looks at the AALs from a service viewpoint. Different Physical Layers may be used by ATM and one of these, the *Synchronous Digital Hierarchy* (SDH), is described in subsection 2.6.1. The SDH is, in itself, an enormous topic and that section only considers ATM aspects; detailed information on SDH can be obtained from many standard textbooks, including [14].

SDH

2.2 Layer-to-layer parameter passing

The B-ISDN parameter passing protocol is based on the ISO OSI model (Figure 2.3). Frames being passed between layers are known as *N-primitive.types* messages, *N* referring to the layer within the model, for example the ATM layer.

Informally speaking, for transmission, each layer takes the frame passed to it from the layer above, adds its own control information and passes it to the layer beneath. On reception, each layer reads and removes the control information added by its peer layer in the transmitting machine and passes the stripped frame upwards.

More formally, frames being transmitted and passing from layer N to a lower layer $N - 1$, are contained in a *Request Primitive* and consist of the user data to be transferred within an N-Service Data Unit (the N-SDU or, equivalently, the (N+1)-PDU) and the Protocol Control Information (PCI) needed between the two N-layer entities. The N-PCI and N-SDU together form the N-PDU which is passed to the lower layer where it is treated as the (N−1)-SDU. The entry point at transmitting layer N is known as the *N-Service Access Point* (N-SAP).

Request Primitive N-SDU PCI

N-SAP

In the other direction, frames passing from a lower layer, $N - 1$, to a higher layer, N (i.e. being received by the computer), are encapsulated in an *Indication Primitive*. The information is an (N−1)-SDU (or, equivalently, an N-PDU) and therefore comprises the user data contained in an N-SDU and the layer N Protocol Control Information. This Protocol Control Information will be stripped at layer N and the remaining N-SDU passed up to layer $N + 1$.

Indication Primitive

Extra information is passed between layers, either through extra parameters in a primitive already in use, or by using further primitives. For example, Quality of Service (QoS) information is contained within extra parameters in the primitive containing the data. Abort information would be indicated using a further primitive, i.e. N-ABORT.indication or N-ABORT.request.

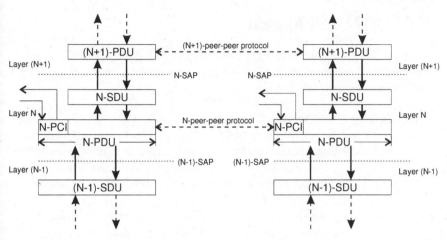

Fig. 2.3 *ISO OSI parameter passing*

2.3   User interface

Two primitives exist at the ATM-SAP, that is at the interface between the ATM layer and the AAL, and are known as the ATM-DATA.request and ATM-DATA.indication. These primitives contain the following information:

- ATM-DATA.indication (cell being received)

  ATM-SDU This field is, naturally, mandatory and contains the 48-octet payload of the ATM cell.

  Congestion indication A mandatory indication of a cell passing through a congested network switch.

  AUU indication This mandatory field allows the transparent communication of user-to-user information.

- ATM-DATA.request (cell being transmitted)

  ATM-SDU This field contains the 48-octet payload of the ATM cell.

  Submitted CLP This field contains the *Cell Loss Priority* (CLP), which *CLP* indicates the relative priority of this cell when cells have to be dropped because of congestion.

  Congestion indication An optional field whose purpose is not specified. It may be required for the support of other services.

  AUU indication The corresponding mandatory field for the ATM-DATA.indication AUU field.

A further Service Access Point (SAP) exists for the OAM interface, known as the *ATM Management-SAP*. This management interface is not described in this *ATMM-SAP* book although the principles and details of the OAM protocols are described in Chapter 5.

2.4 ATM cell format

As described in Chapter 1, an ATM cell has a compromise length of 48 octets of payload and a 5-octet header. The structure of the header is shown in Figure 2.4 and the meanings of the acronyms are as follows:

GFC

GFC. Generic Flow Control. This is only used at the User–Network Interface (UNI) and its function (see I.150 – reference [8]) is to ensure that spare capacity is fairly distributed between users. GFC is used between a Network Terminator and the user's equipment for individual links or for a shared medium, for example a local area network. Figure 2.5 shows the two configurations for GFC implementation.

B-ISDN will eventually support two types of shared medium transmission: uncontrolled transmission, which does not implement any GFC functions, and controlled transmission. Since no GFC procedures have yet been defined, current ATM terminals use uncontrolled transmission. ITU-T recommendation I.361 (reference [15]) differentiates between uncontrolled and controlled transmission as follows:

> 'Any piece of equipment which receives ten or more non-zero GFC fields within 30 000 cell times should consider the other ATM entity to be executing the "controlled transmission" set of procedures.'

If two peer entities implement different transmission procedures, then the layer management is notified by the ATM layer via the ATMM-SAP.

VPI

VPI. Cells are routed along virtual paths through the network. This field, the Virtual Path Identifier, specifies the virtual path being followed by this cell and is negotiated at the UNI and NNI at call initialisation. Note that it is 8 bits long at the UNI and 12 bits long at the NNI, on the assumption that there are likely to be more virtual paths between two network switches than between an end-user and the network. Some VPIs are reserved for special purposes – see section 2.7.

VCI

VCI. Each virtual path is subdivided into a number of channels. This field, the Virtual Channel Identifier, specifies to which virtual channel the cell belongs and, like the VPI, is negotiated at the time the call is established. Also like VPIs, some VCIs are reserved for special purposes – see section 2.7.

PTI

PTI. This field, the Payload Type Indicator, is a 3-bit field that classes the cell payload according to the information which it is carrying: user or OAM information. In earlier versions of the ATM specifications, this field was only 2 bits long and, as can be seen from Table 2.1, it still really consists of a 2-bit field specifying the cell type and a 1-bit field specifying whether the cell met congestion on its journey through the network. This single-

EFCI

bit field, known as the Explicit Forward Congestion Indicator (EFCI), is

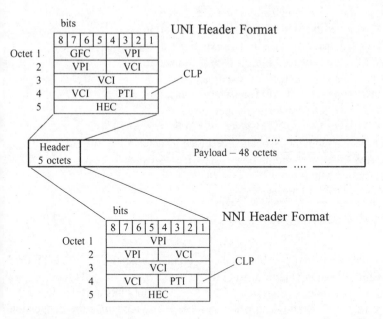

Fig. 2.4 *ATM cell structure*

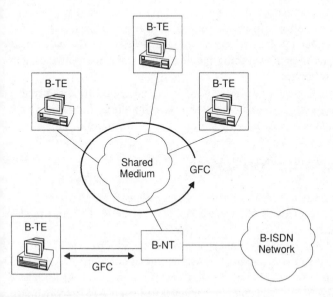

Fig. 2.5 *Shared medium and direct link GFC*

PTI	Meaning
000	User data cell, congestion not experienced ATM-layer-user-to-ATM-layer-user indication = 0
001	User data cell, congestion not experienced ATM-layer-user-to-ATM-layer-user indication = 1
010	User data cell, congestion experienced ATM-layer-user-to-ATM-layer-user indication = 0
011	User data cell, congestion experienced ATM-layer-user-to-ATM-layer-user indication = 1
100	OAM F5 segment-associated cell (see Chapter 5)
101	OAM F5 end-to-end-associated cell (see Chapter 5)
110	Resource Management
111	Reserved for future use

Table 2.1 *PTI values*

optional and its worth relies on the destination, when it receives cells with this bit set, being able to instruct the transmitter to reduce its sending rate and thereby reduce congestion. Clearly the bit only has useful value if the time required for the destination to receive and recognise the congestion bits and instruct the source to reduce its sending rate is significantly less than the time that the congestion lasts.

CLP

CLP. Cell Loss Priority allows for two levels of Quality of Service. Cells with CLP = 1 are assumed by the network to have lower priority than cells with CLP = 0 and are discarded first if congestion occurs. This field may be set either by the user or by the access point to the network if it finds that the user is exceeding the negotiated transmission bandwidth (see I.371 – reference [16]).

HEC

HEC. Header Error Control is an 8-bit CRC value generated by the physical layer. It is used to detect and correct errors within the cell header and for cell delineation: see subsection 2.6.3 on page 34 for more details.

2.5 ATM Adaptation Layer

The ATM Adaptation Layer (AAL) interacts between the different services and the underlying ATM network. I.363 (reference [17]) has defined various modes for the AAL and these are discussed in Chapter 3.

CS
SAR

In general the AAL is divided into a *Convergence Sublayer* (CS) and a *Segmentation And Reassembly Sublayer* (SAR) as shown in Figure 2.6. Service-dependent functions are provided at the AAL-SAR by the CS. As its name suggests, the SAR sublayer segments higher level data into ATM payload size cells at the source and reassembles the original information from the cells at the destination.

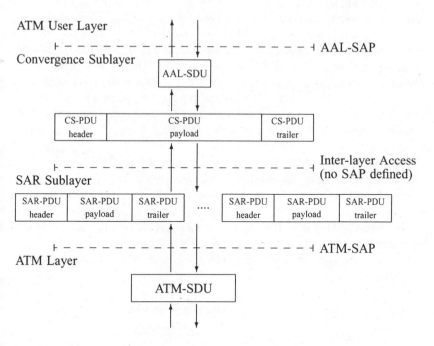

Fig. 2.6 *AAL SAR and CS Sublayers*

2.6 Physical Layer

2.6.1 Transmission scheme

The physical layer may be imagined as a piece of wet string along which the bits of the ATM cells travel. Several possibilities exist for its actual implementation, for example the *Plesiochronous Digital Hierarchy* (PDH) which is already in- *PDH* stalled in most countries and which will allow a smooth introduction of B-ISDN into the existing networks. The proposed physical layer standard is, however, the *Synchronous Digital Hierarchy* (SDH) or *SONET* as its near equivalent is *SDH* known in North America (see I.432 – reference [18]). The SDH and SONET *SONET* standards were published in 1988–1989 and that a telecommunications standard introduced in the late 1980s could differ between North America and the rest of the world is an indictment of the standardisation bodies and a nuisance to the telecommunications industry. For an in-depth discussion of Synchronous Networks see reference [14]; the following is a brief overview of SDH and its terminology.

The SDH protocol transmits information in frames, each frame requiring 125 μs to transmit. A hierarchy of SDH protocols is defined but currently the

STM-1
OC-3

one most commonly used is *STM-1* which is equivalent to OC-3 in the North American SONET standard. See the glossary entry for *plesiochronous* for a table giving the speeds of these links and the relationship between the SDH and SONET naming. Each STM-1 frame comprises $9 \times 270 = 2430$ octets of information and these 2430 octets are normally shown as 9 rows each of 270 octets (columns) as illustrated in Figure 2.7. Octets are transmitted from top-left to bottom-right. The STM-1 frame can be considered, like an ATM cell on a vast scale, to have a header and a payload, but the header, rather than being transmitted before the cell, is transmitted in pieces during the transmission of the cell.

VC-4

The payload of the SDH frame is called a *Virtual Container Level 4* (VC-4) and ATM cells are mapped into this. Unfortunately the length of an ATM cell, 53 octets, does not divide without remainder into the payload length of a VC-4, $9 \times 260 = 2340$ octets, and so an ATM cell has to straddle two consecutive STM-1 frames. This does not cause alignment problems since the ATM cells are aligned to the octet boundaries of the STM-1 frame.

As one frame is transmitted every 125 μs, simple arithmetic shows that the

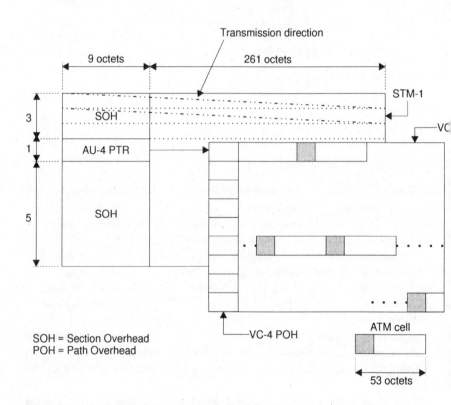

Fig. 2.7 *155.52 Mb/s STM-1 Frame with VC-4*

bit rate of the STM-1 frame is

$$\frac{9 \times 270 \times 8}{125 \times 10^{-6}} = 155.52 \text{ Mb/s}$$

The 9 SDH octets at the beginning of every row and the VC-4 path overhead reduce the payload to 149.76 Mb/s. The mapping of ATM cells, with their 48/53 efficiency, means that the actual bandwidth of an STM-1 link being used to carry user traffic (as distinct from SDH or ATM overhead) is 135.6 Mb/s or 87% of the raw link capacity. When the ATM cells are being used to carry information which is itself encapsulated with headers (for example, Frame Relay traffic), the actual user traffic throughput drops further.

The next transmission speed in the SDH hierarchy is STM-4 (OC-12 in North America) and this frame is shown in Figure 2.8.

2.6.2 Physical Layer cells

It is necessary for the Physical Layer to detect the beginning of a cell when a link first becomes active. This process, known as *cell delineation*, is complicated by the SDH frame offsets and the fact that an STM-1 frame does not contain a whole number of ATM cells. Once one cell's beginning is found and confirmed, the Physical Layer can detect the beginning of successive cells simply by counting batches of 53 octets although it still needs to monitor what it is finding to ensure that it is not being confused by a transmission error. To find the alignment of the ATM cells, the Physical Layer uses the Header Error Control (HEC) – see *HEC* section 2.6.3 – extracting these fields from the SDH frame. The physical layer determines the type of the cell by interrogating the ATM cell header. From a physical layer view point, cells are classed into the following types:

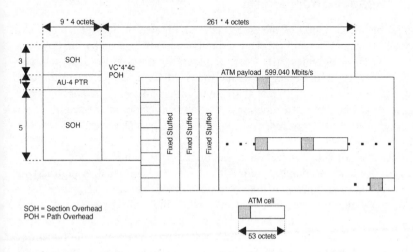

Fig. 2.8 *622.08 Mb/s STM-4 Frame With VC-4-4c*

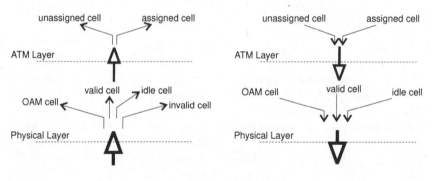

Fig. 2.9 *Physical Layer handling of cells*

- Idle cell. Such a cell is generated and removed at the Physical Layer to maintain the correct cell flow rate and is not seen by higher layers.
- OAM cell. This type of cell is directed by the Physical Layer to or from the OAM functions.
- Valid cell. A cell at the Physical Layer that has no header errors or only errors that have been corrected by the header error control (HEC). Such a cell is passed to or from the ATM layer.
- Invalid cell. A cell at the Physical Layer that has header errors which cannot be corrected by the Physical Layer. These cells are counted and discarded.

The handling of these cells by the Physical Layer is illustrated in Figure 2.9.

2.6.3 Header Error Control

Although the HEC is part of the ATM cell, see Figure 2.4, it is generated and inserted on transmission and checked and removed on reception by the Physical Layer. The HEC provides error correction/detection and allows the physical layer to detect the cell boundaries within the STM-1 frame. Detecting the cell *cell* boundaries in what is initially just a stream of bits is known as *cell delineation*. *delineation* Thus the HEC performs two functions: it allows bit errors in the header to be detected and possibly corrected, and it allows the Physical Layer to detect the ATM cell boundaries.

Header error correction/detection works on a 2-state model as shown in Figure 2.10. Because the number of bits used for the HEC field is high (8) in comparison with the number of bits being protected ($4 \times 8 = 32$), some error correction can also be performed, but only with single bit errors. Initially the system is in the mode where it will correct single bit errors and it will stay in this mode until an error of some form is detected. When multiple bit errors occur the probability of detecting the error in the correction state is substantially lower than in the detection state. Therefore, when any error is detected in the

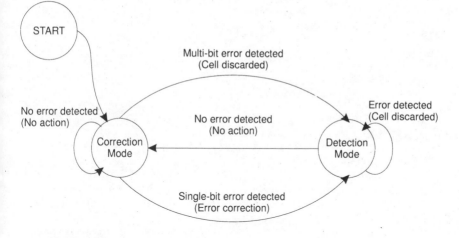

Fig. 2.10 *HEC error correction and detection*

correction state, the HEC algorithm switches to the detection state. The HEC algorithm switches back to the correction state only when an error free cell has been received.

The HEC field is an 8-bit Cyclic Redundancy Check (CRC) Code with gen- *CRC* erator polynomial $x^8 + x^2 + x + 1$. To calculate the CRC code, the first 4 octets of the header are multiplied (modulo 2) by x^8. This result provides the denominator which is then divided (modulo 2) by the generator polynomial. The remainder is then added (modulo 2) to the denominator. This sounds complex but CRC algorithms have been designed to be easily realised in hardware. For more details of the CRC algorithms, see any text on communications theory, including reference [19].

ITU-T recommendation I.432 (reference [18]) defines the mechanism to identify cell boundaries within the STM-1 frame. The basic technique is to hunt for a correct CRC, moving along the bit stream until one is found. Even then it cannot be guaranteed that the correct start-of-cell has been found as the user information inside the cells might be such that it gives a correct CRC at the wrong point purely by chance. More formally stated, a 3-state algorithm, as shown in Figure 2.11, begins by monitoring the unsynchronised incoming bit stream and performs CRC error detection with the HEC polynomial until no more errors are found. This is known as the HUNT state. When the CRC is valid then it is highly probable, but not certain, that the cell header has been detected. The algorithm then switches into the PRESYNCH state where it performs HEC error detection on a cell-by-cell basis. If the HEC fails then the algorithm switches back to the HUNT state on the assumption that it found a false start-of-cell. Provided the PRESYNCH state correctly identifies δ consecutive HECs then the system is synchronised and moves to the SYNCH state.

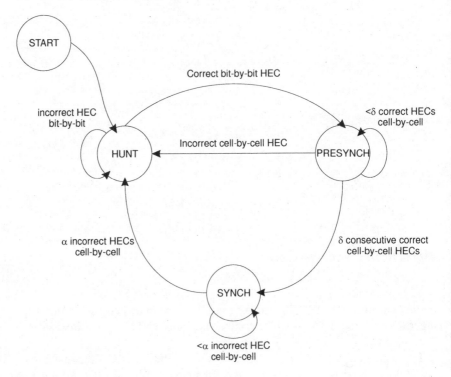

Fig. 2.11 *HEC delineation*

Until synchronisation is lost, the normal 2-state HEC algorithm applies. Lost synchronisation is determined by α consecutive incorrect HECs. The values of $\alpha = 7$ and $\delta = 6$ have been suggested for SDH-based interfaces.

Cell delineation could still find an incorrect start-of-cell if an application or malicious user implemented the same generator polynomial in the user data. To overcome these possibilities, two scramblers have been recommended. For SDH-based Physical Layer systems a self synchronising scrambler is implemented at the physical layer. The scrambling polynomial $x^{48}+1$, is implemented over the information field only and is descrambled during the PRESYNCH and SYNCH states. This is in conjunction with any other techniques used at the transmission protocol level.

PRBS   For cell-based physical layer systems a 31st order distributed scrambler is recommended. A pseudorandom binary (PRBS) sequence is generated by the polynomial $x^{31} + x^{28} + 1$. This is added (modulo 2) to the entire cell with the exception of the HEC field. The HEC has the most significant 2 bits modified by addition (modulo 2) of:

- HEC bit 8 and the Pseudorandom Binary Sequence 211 bits earlier, i.e. half a cell apart

- HEC bit 7 and the current PRBS.

During cell delineation only the last six bits of the HEC can be used since the most significant 2 bits were modified and cannot be evaluated until the scrambler is synchronised. Scrambler synchronisation is achieved by using the two most significant HEC bits that are linearly independent. Thirty-one sample bits conveyed by bits 7 and 8 of the HEC are required to synchronise, that is 16 cells. Detection of the scrambler's presence is achieved by monitoring bits 7 and 8 for the PRBS equal to zero in all cases. For further information see I.432 – reference [18].

This randomising technique will be successful even against the most determined malicious user. Since the user cannot determine precisely where in the STM-1 frames her cells will be transmitted, she cannot fool the transmission equipment into falsely synchronising by putting special values into her own data.

2.7 Pre-assigned cell headers

I.361 (reference [15]) defines pre-assigned cell headers for use at the physical and ATM layers of the UNI, see Tables 2.2 and 2.3. In the tables, the terms F1 to F5 refer to Operation and Maintenance (OAM) functions that are described in Chapter 5. Comparing the two tables it can be seen that physical layer cells and unassigned cells are distinguished by the CLP bit, preventing the use of the cell loss priority scheme described on page 30 on physical layer or unassigned cells (see I.150 – reference [8]).

Use	VPI	VCI	PT	CLP	HEC
Idle Cell Identification [20]	00 (hex)	00 (hex)	000	1	Valid
Physical Layer OAM F3 Flow Cell [20]	00 (hex)	00 (hex)	001	1	01011100
Physical Layer OAM F1 Flow Cell [20]	00 (hex)	00 (hex)	100	1	01101010
Reserved for use of Physical Layer [18]	00 (hex)	00 (hex)	PPP	1	Valid
Note:	P indicates the bit is available for use by the physical layer. This field is independent of the ATM layer bits				

Table 2.2 *NNI Pre-assigned header values*

Use	VPI	VCI	PT	CLP	HEC
Meta-signalling [20]	note 1	00000000 00000001 (note 5)	0A0	C	Valid
General Broadcast Signalling [20]	note 1	00000000 00000010 (note 5)	0AA	C	Valid
Point-to-point signalling [12]	note 1	00000000 00000101 (note 5)	0AA	C	Valid
Segment OAM F4 flow cell [12]	note 2	00000000 00000011 (note 4)	0A0	A	Valid
End-to-end OAM F4 flow cell [12]	note 2	00000000 00000100 (note 4)	0A0	A	Valid
Segment OAM F5 flow cell [12]	note 2	zzzzzzzz zzzzzzzz (note 3)	100	A	Valid
End-to-end OAM F5 flow cell [12]	note 2	zzzzzzzz zzzzzzzz (note 3)	101	A	Valid
Resource Management [16]	note 2	zzzzzzzz zzzzzzzz (note 3)	110	A	Valid
Unassigned cell	00000000	00000000 00000000	BBB	0	Valid
Key	A implies that a bit may be 0 or 1 as set by the appropriate ATM layer function B indicates a *don't care* bit C indicates that the bit is available for use by the physical layer				
Note 1	VPI = 0 only for user signalling with the local exchange. VPI other than zero is used for signalling with other entities				
Note 2	Any VPI value other than zero				
Note 3	Bits are available for the GFC protocol				
Note 4	The VCI values are pre-assigned in every VPC at the UNI.				

Table 2.3 *UNI pre-assigned header values*

3
B-ISDN SERVICE INTERFACE

3.1 Introduction

As discussed in Section 2.1, the interface between the service and B-ISDN is known as the ATM Adaptation Layer (AAL). A protocol exists (or will exist) to implement each of the service classes defined in Table 1.2 as follows:

- AAL type 0 (AAL-0), a common way of referring to a direct transfer of data with no adaptation
- AAL type 1 (AAL-1), for constant bit rate services
- AAL type 2 (AAL-2), ostensibly for variable bit rate services but, since this has never been defined, the term is now being reused as indicated on page 15, to define a protocol that carries multiple CPS-Packets per cell
- AAL type 3/4 (AAL-3/4), for connection-oriented and connectionless data transfer services
- AAL type 5 (AAL-5), for connection-oriented data services.

In brief, the AAL-1 must deal with synchronisation, timing and structured data transfer for real-time CBR services. Consequently timing bits and pointers have to be carried to match local clocks to the network clock and to define where data structures begin and end. Since retransmission is particularly detrimental to real-time applications, a Reed–Solomon Code (128, 124) is used to increase the probability of recovering a corrupted message without retransmission.

AAL-2 in its original form has not been developed; details of the original requirements are given in section 3.3. The 'new' AAL-2 as defined in I.363.2 (reference [21]) is expected to support low data rate applications with multiple packets stored in each cell.

AAL-3/4 was designed to support time-insensitive data transfer. Consequently the protocol concentrates on reliable data transmission and additional multiplexing facilities. It has enhanced capabilities for identifying misordered cells, the beginning and end of messages and message lengths as well as cell-based CRC checking. In addition, a multiplexing field allows multiple data streams to be defined for each channel. AAL-3/4 has largely been superseded by AAL-5 due to the large overhead incurred in the AAL-3/4 cell structure.

AAL-5 represents a minimum overhead approach to data transmission. All 48 octets of the SAR PDU are used for the SAR SDU, error support is done at frame level with a large CRC and length field. Since modern networks have low error rates the more frequent retransmissions required by the 'lighter' protocol are regarded as acceptable. Unlike AAL-3/4 it is limited to 65 535 octets per CPCS-PDU and it mixes levels by using the cell header to carry user information.

The following sections discuss the AALs in more detail. Several service types may be supported by one AAL and additional information structuring may therefore be required. These additional protocols for connectionless and Frame Relay services are discussed in sections 3.5 and 6.4.

Further information can be found in recommendation I.363 (reference [17]). The ability to create new AALs allows for later developments to be incorporated into B-ISDN. Section 3.6 outlines proposed AALs from non-ITU-T sources.

3.2 AAL-1 and Class A (CBR)

3.2.1 User interface

The functions provided by AAL-1 reside in the Convergence Sublayer (CS) and are accessed through the AAL-SAP using the following primitives:

- AAL-UNITDATA.indication containing a mandatory data field and an optional structure field.

  This primitive is sent by the AAL to the layer above it to notify the receiving layer that a cell has arrived. The size of the Data parameter remains constant and is determined by the service provided. The request is sent at a predetermined rate. The Structure parameter is discussed later.
- AAL-UNITDATA.request containing a mandatory data field and an optional structure field.

AAL-SDU information contained in the data parameter is delivered to the AAL user using the indication. Both data parameter size and time interval are provided for by the AAL. For video and voice, the AAL-SDU Data parameter is one octet wide, although for circuit transport of asynchronous circuits using the Synchronous Residual Time Stamp (SRTS) method, see section 3.2.3, it is only one bit wide (see I.363 − reference [17]).

SRTS

Both plesiochronous circuit transport, for example constant bit rate signals at 1544, 2048, 6312, etc. kb/s, and synchronous transport, for example constant bit rate signals at 64, 384, 1563, etc. kb/s, are supported by AAL-1.

In both the request and indication primitives, the Structure parameter can hold the values *Start* and *Continuation* and its usage is agreed upon before connection. The indication primitive can also use the Status parameter to indicate *Valid Data* and *Invalid Data*. The use of this parameter is agreed before connection of the AAL user and AAL-1.

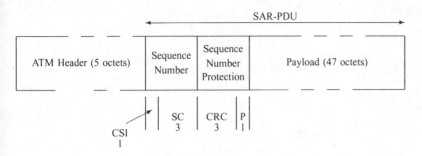

Fig. 3.1 *SAR-PDU cell structure for AAL-1*

3.2.2 Cell structure

AAL-1 allows a maximum of 47 of the 48 octets in the ATM-SDU to be used
as SAR-PDU payload. The remaining octet implements the SAR-PDU header.
The header is divided into two: a 4-bit Sequence Number (SN) field and a *SN*
4-bit Sequence Number Protection (SNP) field, as shown in Figure 3.1. The SN *SNP*
is further subdivided into a 3-bit Sequence Count (SC) field and a 1-bit Con- *SC*
vergence Sublayer Indication (CSI) field. The SC is a simple 3-bit (modulo 8) *CSI*
count of cells and detects whether a cell is in its correct sequence, whether a cell
has been lost or whether a cell has been misinserted. This 3-bit count provides
poor protection in systems with a high cell loss, that is a cell loss probability
greater than 10^{-6} per cell (see reference [3]).

The CSI bit defines the use of timing information or structured data, depending
on the service.

Sequence numbering is protected by the Sequence Number Protection (SNP)
field which is subdivided into a 3-bit CRC of the SN field and an even parity bit calculated over the 7 bits of the SN, CSI and SNP. A 2-state error
correction/error detection algorithm is used (see Figure 3.2).

3.2.3 Service implementation

The convergence sublayer has been standardised to provide for CBR circuit,
video signal and voice data. Structured data services require the bits of the data
stream to be organised into groups, generally of a multiple of 8 bits, and for the
boundaries of the groups to be delineated. The AAL STRUCTURE parameter option within the AAL-UNITDATA.request and AAL-UNITDATA.indication
can be used to implement Structured Data Transfer (SDT). A pointer is imple- *SDT*
mented to delineate the structure boundaries. SAR-PDU payloads containing
the pointer value are denoted as P format; non-P format indicates all other *P and non-P*
SAR-PDU payloads. *formats*

The CSI bit is set to TRUE if structured mode is in use, and in that particular

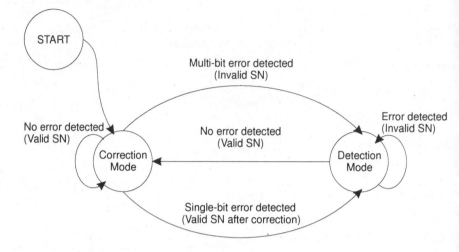

Fig. 3.2 *SNP error correction/detection algorithm*

SAR-PDU, there is a second overhead byte leaving only 46 bytes for payload. This supplementary overhead contains a pointer which holds an offset, in octets, between the end of the pointer field and the first octet of the next structured block. The payload is constructed from the 46 octets of the P format SAR-PDU payload and the 47 octets of the next SAR-PDU payload, non-P format, thus requiring a maximum pointer value of 93.

For services that require timing information, such as video conferencing, two approaches may be taken, the Synchronous Residual Time Stamp method (SRTS) or Adaptive Clock method. The most effective means of source clock frequency *SRTS* recovery uses the SRTS method. A 4-bit Residual Time Stamp is carried, one bit per cell, by the CSI bit in cells with ODD sequence count values, that is sequence counts 1, 3, 5, 7 (note that the sequence count is a 3-bit number). The residual time stamp indicates the frequency difference between a common network clock and a local service clock.

Adaptive The adaptive clock method maintains a buffer which is read at a frequency *Clock* determined by a local clock. If the buffer is nearly full, then the clock must be running at too low a frequency and the frequency is adjusted upwards. If the buffer is nearly empty the converse happens. This method assumes that there will be no sudden changes in the frequency of the network clock.

SDT and SRTS share the CSI bit. As above, the Residual Time Stamp (RTS) is implemented in odd sequence count values. These are considered to be non-P format SAR-PDU payloads. Even sequence count values (that is sequence counts 0, 2, 4, 6) always contain P format SAR-PDU payloads. A pointer value of 127 indicates that no structure boundary falls within the next 93 octets. In the example shown in Figure 3.3, which illustrates this technique, block one is assumed to have a size of 230 octets. The beginning of block 1 falls part-

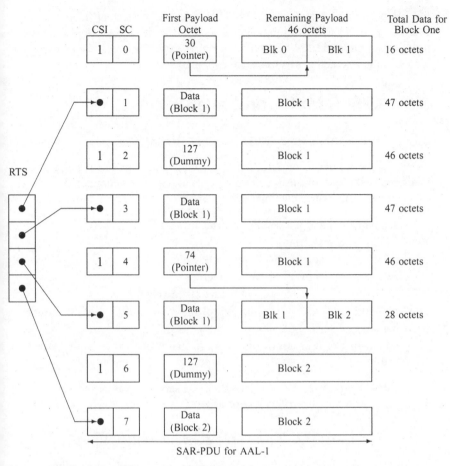

SAR-PDU for AAL-1

Note: example shows 230 octets in block one

Fig. 3.3 *Structured data transfer using SRTS*

way through the cell with sequence number 0. A dummy pointer, having the value 127, indicates that the block continues beyond the next two cells. Note that the CSI bit is always TRUE on an even sequence count.

Cell delay variation, an inherent ATM characteristic and cause of distortion for audio processing, is removed by buffering at the receiver. The size of the buffer is not contained in the standards and is the subject of much research. However, for instances of buffer overflow and underflow the CS is responsible for the insertion and removal of bits to maintain a constant flow.

Bit error correction and cell loss detection for unidirectional video services are achieved by using Forward Error Checking (FEC) and octet interleaving. The *FEC*

FEC is a Reed–Solomon (128, 124) Code and can correct up to two errored octets or four deleted octets in one 128 octet block (see reference [4]). In order to construct this code 124 cell delays are necessary. This delay is unacceptable for real-time interactive services such as voice.

3.3 ATM Adaptation Layer-2

3.3.1 Original AAL-2 and Class B (VBR)

As was discussed in section 3.1, AAL-2 has undergone significant changes since first proposed. The original definition of AAL-2, never completed in detail, called for the following functions to be performed (see I.363 – reference [17]):

- segmentation and reassembly of user information
- handling of cell delay variation
- handling of lost and misinserted cells
- source clock recovery at the receiver
- recovery of the source data structure at the receiver
- monitoring of AAL-Protocol Control Information (PCI) for bit errors
- handling of AAL-PCI bit errors
- monitoring of user information field for bit errors and possible corrective action.

In practice, this AAL for Variable Bit Rate services was never defined and the AAL-2 term now describes a very different adaptation layer.

3.3.2 Revised AAL-2 (Multiple packets per cell)

The revised version of AAL-2 is being developed to allow the user to have multiple AAL connections per ATM connection. This is achieved by multiplexing 'packets' of data in the Common Part Sublayer (CPS). The Quality of Service (QoS) is set separately for each AAL-SAP, but significant work still needs to be done in this area. For example, the mapping of the QoS to the ATM-SAP is still under investigation along with the problem of how to implement point-to-multipoint connections. A number of packets can be placed within an ATM cell.

CPS Packet   The default length for user data in a CPS packet is up to 45 octets – with the three-octet CPS packet header, this guarantees that the full CPS packet will be no longer than 48 octets and can fit into a single ATM cell. The CPS Packet Header (CPS-PH) has a length indicator of six bits, and the user can specify a maximum length of 64 octets if required. The CPS-PDU may contain complete or partial CPS-Packets. In addition to the Length Indicator (LI), the CPS-PH contains a Channel Identifier (CID), User-to-User Indentification (UUI) and Header Error Control (HEC), as shown in Figure 3.4.

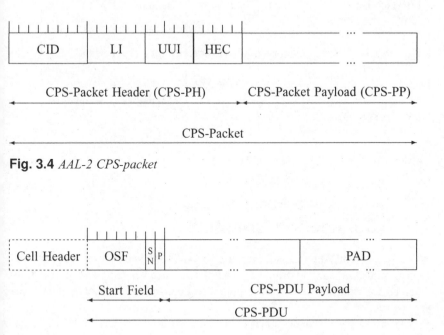

Fig. 3.4 *AAL-2 CPS-packet*

Fig. 3.5 *AAL-2 CPS PDU*

The channel identifier does not use 0x00, as this is used for padding, or 0x01 to 0x07 which are reserved for layer management peer-to-peer procedures. This leaves 0x08 to 0xFF as user identifiers. Five bits are used for User-to-User Identification of which the codes 0x00 to 0x1B are used for SSCS entities, 0x1E and 0x1F are for layer management and the remaining two are reserved for later use. Finally, the HEC is a standard CRC with generator polynomial $x^5 + x^2 + 1$ which covers the rest of the packet header.

These packets are multiplexed into the ATM-cells. The CPS-PDU, which *CPS-PDU* corresponds to the ATM-SDU, has a 47-octet payload with a single-octet start field containing a six-bit offset field (OSF), a sequence number bit (SN) and an odd parity bit (P). If the CPS-PDU is too large to be filled by the CPS packets, zeros are used to pad the payload as shown in Figure 3.5.

The UNITDATA primitives for this AAL-2 are shown in Table 3.1.

This AAL is expected to be under discussion until the year 2000, although implementations are already beginning to appear in commercial products. For more details the reader is referred to I.363.2 (reference [21]).

	CPS-UNITDATA		MAAL-UNITDATA	
	REQ.	IND.	REQ.	IND.
Interface Data (CPS - INFO)	Mandatory	Mandatory	Mandatory	Mandatory
User to User (CPS - UUI)	Mandatory	Mandatory	Mandatory	Mandatory
Channel Identifier (CID)	—	—	Mandatory	Mandatory

Table 3.1 *UNITDATA primitives for AAL-2*

3.4 AAL-3/4, 5 and Class C (Connection-Oriented)

ATM is inherently connection-oriented so AAL-3/4 and AAL-5 have a relatively simple and similar structure. AAL-3/4 has been specified for connection-oriented data transfer, as discussed in this section; for connectionless data transfer see section 3.5.

AAL-5 has only been specified for connection-oriented data transfer and has effectively been developed as a simpler replacement for AAL-3/4 which is really too heavy and complex for data transfer. Since the objectives of both AALs are the same they have a similar architecture. The significant differences between AAL-3/4 and AAL-5 are that:

- AAL-3/4 offers greater flexibility, but incurs a greater overhead.
- AAL-5 provides greater error protection over its frame structure and implements a cleaner architecture (see section 3.4.3).

SAR
CS

SSCS
CPCS

AAL-3/4 and AAL-5 have the same layered architecture consisting of two main sublayers known as the Segmentation and Reassembly Sublayer (SAR) and the Convergence Sublayer (CS), see Figure 3.6. To allow the input data stream to enter the AAL in several different ways, the convergence sublayer has been divided into two further sublayers. These sublayers divide the task of data handling into service-specific and common data functions. The sublayers are called the Service-Specific Convergence Sublayer (SSCS) and the Common Part Convergence Sublayer (CPCS). The SSCS has not been defined for either AAL. The recommendations state that it may do nothing so that it is transparent and all data handling is performed by the CPCS.

Subsection 3.4.1 discusses functions common to AAL-3/4 and AAL-5. Functions specific to AAL-3/4 and AAL-5 are discussed in subsections 3.4.2 and 3.4.3 respectively.

3.4.1 AAL-3/4 and AAL-5 common functions

Message and Streaming Modes

AAL-3/4 and AAL-5 provide two modes of user data input known as *Message Mode* and *Streaming Mode*. Message Mode is used to convey information which *Message* is already in packet format (for example, Frame Relay data) whereas Streaming *Mode* Mode is used for low-speed continuous data, often with low delay requirements. *Streaming*

In message mode the service provides for complete AAL-SDU input, whether *Mode* small fixed size or variable length. The SSCS is expected to provide the functions to translate AAL-SDUs into SSCS-PDUs by means of concatenation, segmentation or direct transport; see Figure 3.7. Note that the AAL-SDU must be complete before SSCS functions are performed. The SSCS-PDU, and likewise the CPCS-SDU, are passed across the interface in exactly one CPCS-interface data unit (CPCS-IDU). The CPCS provides the transport of the CPCS-SDU in one CPCS-PDU. Segmentation and reassembly functions are performed on the CPCS-PDU at the SAR sublayer. The resulting SAR-PDUs are mapped directly into ATM-SDUs and thereby into ATM-PDUs.

Streaming mode allows an AAL-SDU to be passed into one or more AAL-IDUs at the AAL-SAR; this translation may be disjointed in time. All AAL-IDUs belonging to one AAL-SDU are transported by the SSCS by means of concatenation, segmentation or pipelining, the resultant SSCS-PDU is mapped into one CPCS-SDU. The CPCS-SDU is mapped into one or more CPCS-IDUs that may occur at irregular time intervals. All CPCS-IDUs belonging to one CPCS-SDU are transported in one CPCS-PDU; see Figure 3.7. At any time during the transmission the abort primitive can be sent to prevent any more PDUs, including the current, being transmitted.

The above modes of service offer the following quality options.

- *Assured operation* provides retransmission of data at the SSCS-PDU level. This ensures that the AAL-SDU received is the same as the one sent, regardless of errors. This will probably be limited to point–point connections only.
- *Non-assured operation* does not provide for retransmission if AAL-SDUs become lost or corrupted. Corrupt AAL-SDUs may be delivered to the user if required, thereby giving the user freedom over error correction and later retransmission.

User interface

The function of the SSCS layer and the generation of the AAL-UNITDATA primitives in the following discussion are speculative, though one SSCS has been defined for Frame Relay services (FR-SSCS): see subsection 6.4.1. For the purposes of this discussion the SSCS will be ignored. The retention of AAL-UNITDATA primitives makes these AALs consistent with the others. Data can

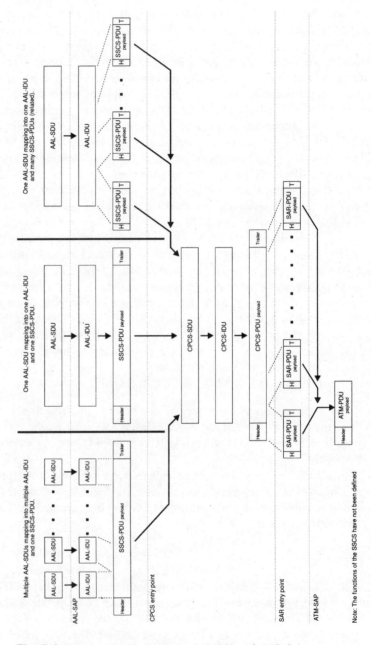

Fig. 3.6 *Message Mode Service: AAL-3/4 and AAL-5*

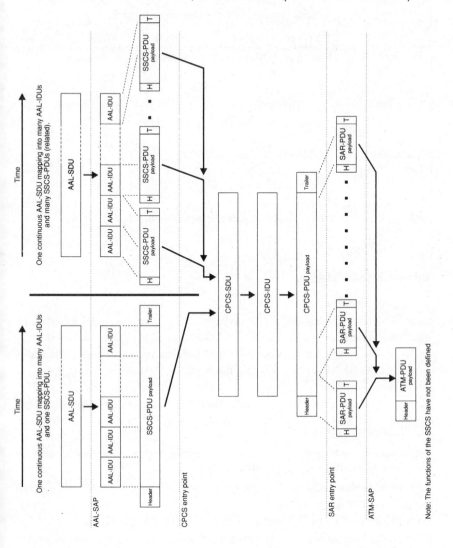

Fig. 3.7 *Streaming Mode Service: AAL-3/4 and AAL-5*

be passed to the AAL via the AAL-SAP using the AAL-UNITDATA-request. This is passed through the SSCS to a CPCS-UNITDATA-invoke primitive. From the CPCS sublayer a CPCS-UNITDATA-signal is translated by the SSCS to an AAL-UNITDATA.indication. The structure of the CPCS-UNITDATA.invoke and CPCS-UNITDATA.signal for AAL-3/4 are shown in Table 3.2 and for AAL-5 in Table 3.3.

The fields in these two tables have the following meaning:

	CPCS-UNITDATA		SAR-UNITDATA	
	INVOKE	SIGNAL	INVOKE	SIGNAL
Interface Data (ID)	Mandatory	Mandatory	Mandatory	Mandatory
More (M)	Mandatory in streaming mode		Mandatory	Mandatory
Maximum Length (ML)	Mandatory in streaming mode with 1st primitive	Optional	—	—
Reception Status (RS)	Optional	Optional	Optional	Optional

Table 3.2 *UNITDATA primitives for AAL-3/4*

Interface Data. This is the actual information to be exchanged across the appropriate (CPCS/SSCS or SAR/CPCS) boundary. For information being passed across the CPCS/SSCS boundary in message mode, the entire SSCS-SDU is contained within the Interface Data. In streaming mode the information may be conveyed in one or more SSCS-SDUs.

More. This field is set to 1 to indicate that the information in the Interface Data does not comprise a complete packet: more data are to come (streaming mode only).

Maximum Length. This is only used in streaming mode and indicates the maximum length of the entire CPCS-SDU.

Reception Status. This field gives the receiver an indication that the current message may be corrupt.

Loss Priority. This field is set if the priority of the packet is low.

Congestion Indication. This field is set if the SDU has experienced congestion on its journey.

User-to-User Indication. This field conveys information between peer CPCS sublayers.

At the SAR/CPCS interface an SAR-UNITDATA-invoke primitive is passed from the CPCS sublayer to the SAR sublayer. A CPCS-UNITDATA-signal is used in the return direction. The structure of the SAR-UNITDATA-invoke and SAR-UNITDATA-signal for AAL-3/4 is shown in Table 3.2, and for AAL-5 in Table 3.3.

Further AAL primitives at the AAL-SAP will be defined at the same time as the SSCS. Note the use of *invoke* and *signal* instead of *request* and *indication*;

| | CPCS-UNITDATA | | SAR-UNITDATA | |
	INVOKE	SIGNAL	INVOKE	SIGNAL
Interface Data (ID)	Mandatory	Mandatory	Mandatory	Mandatory
More (M)	Mandatory (str. mode)	Mandatory (str. mode)	Mandatory	Mandatory
Loss Priority (LP)	Mandatory (note 1)	Mandatory (note 2)	Mandatory	Mandatory
Congestion Indication (CI)	Mandatory (note 1)	Mandatory (note 2)	Mandatory	Mandatory
User-to-User Indication (UU)	Mandatory (msg. mode) (note 2)	Mandatory (msg. mode) (note 2)	—	—
Reception Status (RS)	—	Mandatory (msg. mode)	—	—
Note 1: Mandatory with first INVOKE Note 2: Mandatory with last INVOKE or SIGNAL				

Table 3.3 *UNITDATA primitives for AAL-5*

this is used to show the absence of a SAP between sublayers.

Connections already established can be aborted by using the CPCS-U-ABORT-invoke and CPCS-U-ABORT-signal primitives at the CPCS layer. SAR-U-ABORT-invoke and SAR-U-ABORT-signal are used at the SAR layer. This aborts all future transmissions and the one in progress allowing re-transmission to begin as soon as possible. Figure 3.8 shows the translation of primitives for AAL-3/4 and AAL-5.

3.4.2   AAL-3/4 coding

Segmentation and Reassembly cell coding

The *cell* structure SAR-PDU can be seen in Figure 3.9 and consists of a 2-octet header, 2-octet trailer and a 44-octet payload. The header consists of a Segment Type (ST) field, a Sequence Number (SN) field and a Multiplex Identification *ST* (MID) field. *SN*

Several different classes of SAR-PDU can exist and are determined by the ST *MID* field. A Beginning Of Message (BOM) indicates that the segment is the start of *BOM*

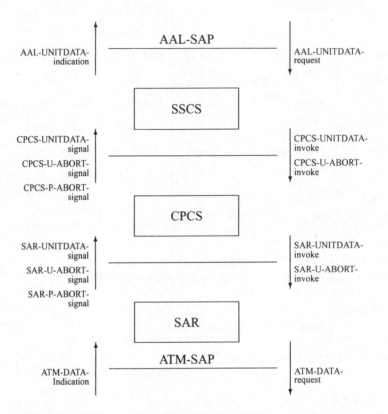

Fig. 3.8 *Primitives within AAL-3/4 and AAL-5*

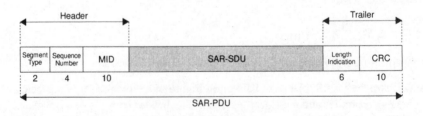

Fig. 3.9 *SAR-PDU frame structure*

a new CPCS-PDU. Messages longer than two SAR-SDUs use a Continuation Of Message (COM) indication in the ST field. COM indicates that the current *COM* segment is a continuation of the current CPCS-PDU. An End Of Message (EOM) *EOM* indicates that the segment is the last in the current CPCS-PDU. Provided the CPCS-PDU is correct the CPCS-SDU can then be passed to the SSCS sublayer. If the entire CPCS-PDU fits into one SAR-SDU then the ST field indicates a Single Segment Message.

The 4-bit SN field is used to detect misordered, misinserted or lost cells within a CPCS-PDU. It is incremented (modulo 16) for each SAR-PDU belonging to the same CPCS-PDU, that is having the same MID value. The current maximum size of the CPCS-PDU is 65 535 octets, that is 1490 SAR-PDUs. The 4-bit SN field will fail to detect sequence errors due to misordering, or misinsertion with a maximum probability of 10^{-8} per cell (see reference [3]).

To allow multiple connections over a single ATM VCC, for example separate data channels for different applications, multiple AAL-3/4 data streams can be created. Each connection within the same VCC is identified by the MID field and has different CPCS and SSCS sublayers. The number of linked messages is determined at call establishment although dynamic handling is also being studied. For connection-oriented services this will be limited to a user–user basis only. The SAR sublayer maintains data integrity at all times. Figure 3.10 shows how multiple SSCSs and CPCSs act independently for each multiplexed link; note that there is still only one AAL-SAP.

Point-to-point or point-to-multipoint services can be provided by the AAL at the AAL-SDU transmit side to the AAL-SAP receive side. The ability to create broadcast messages at this level avoids a requirement for multiple ATM connection for a broadcast signal and is an essential element of efficient transfer of IP over ATM (see Chapter 6).

The SAR-PDU trailer contains two fields, as shown in Figure 3.9. The 6-bit Length Indication (LI) field determines the number of octets present in the *LI* SAR-SDU. BOM and COM SAP-PDUs contain 44 octets and therefore have $LI = 44$. For EOM SAR-PDUs, $4 < LI < 44$, if $LI = 63$ then the transfer is aborted. SSM SAR-PDUs have an LI value satisfying $8 < LI < 44$.

Error detection of the entire SAR-PDU is covered by a 10-bit CRC. The CRC is the remainder of the product of the entire SAR-PDU except the CRC and x^{10}, divided by the generator polynomial $(x^{10} + x^9 + x^4 + x + 1)$.

Convergence Sublayer frame coding

Figure 3.11 shows the CPCS-PDU with a 4-octet header and trailer. The header contains a 1-octet Common Part Indicator (CPI), a 1-octet Beginning tag (Btag) *CPI* and a 2-octet Buffer Allocation Size (BASize). The trailer consists of a 1-octet Alignment (Al), 1-octet End tag (Etag) and a 2-octet Length field. The Al and *Al* PAD fields are used to align the CPCS-SDU to a 32-bit boundary: the Al field aligning the data bits to an octet boundary with zeros and the PAD field aligning

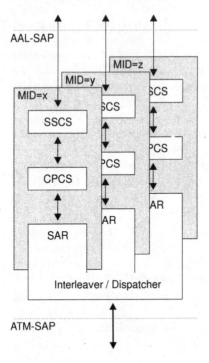

AAL-SAP

MID=z

MID=y

MID=x

SSCS

SCS

SCS

PCS

CPCS

PCS

SAR

AR

AR

Interleaver / Dispatcher

ATM-SAP

Note: Number of multiplexed links established at call connection

Fig. 3.10 *SSCS and CPCS Multiplexing at the SAR Sublayer*

the CPCS-SDU to an integral multiple of four octets. This allows bit streams to be used across the network without the upper layers having to perform alignment functions.

Further error protection by the CPCS-PDU attempts to detect errors due to lost cells and misordering undetected by the SAR sublayer. The values in the *Btag* and *Etag* fields must be equal for one CPCS-PDU. The generation of the tag fields is not specified; simple increments between successive CPCS-PDUs are sufficient. The CPCS sublayer does not check sequencing between successive CPCS-PDUs. The Length field stores the length of the entire CPCS-PDU, the unit of the Length field being determined by the CPI. Currently only octets have been defined as valid lengths and the maximum CPCS-PDU size is $2^{16} - 1 = 65\,535$ octets. Allowing CPCS-PDUs greater than 65 535 octets makes the protocol future-proof.

Btag
Etag

The receiver can pre-allocate a sufficient buffer for the frame as soon as the BASize field has been received, as this field contains the size of the entire CPCS-PDU in units defined by the CPI field. This field can be extracted from the first SAR-PDU received and the CPCS layer can then allocate the necessary

BASize

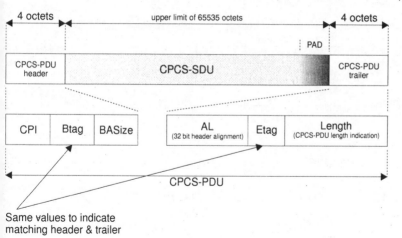

Fig. 3.11 *CPCS-PDU frame structure for AAL-3/4*

buffer space. A quick error check can be performed by comparing the Length field and the BASize field, which should both be the same.

3.4.3 AAL-5 coding

Segmentation and Reassembly cell coding

AAL-5 allows all 48 octets of the SAR-PDU to be used for the SAR-SDU, that is for payload data. This gives a 10% increase in the amount of data transmitted over the network in one SAR-PDU. Multiplexing functions do not exist within AAL-5 and multiple connections are achieved at the ATM layer. To determine if the data being received are the last SAR-PDU in a CPCS-PDU the payload type field of the ATM layer is monitored. The ATM-layer-user to ATM-layer-user (AUU) parameter in the ATM layer primitive relays the data to the AAL. *AUU* The AUU parameter is transported across the network by the PT field in the ATM cell header; see Figure 3.12. If the PT field is 0 then the SAR-PDU is a Beginning Of Message (BOM) or Continuation Of Message (COM). A PT field of 1 indicates either an End Of Message (EOM) or a Single Segment Message (SSM). Using the PT field introduces *level mixing* which does not comply with the ISO OSI model although AAL-5 is widely accepted by ATM equipment suppliers and their customers.

The CPCS sublayer incorporates padding to fill the CPCS-SDU to the SAR-PDU 48-octet boundary, (see Figure 3.11). This eliminates the need for a field in the SAR-PDU to determine the number of octets in the SAR-PDU.

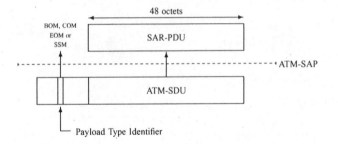

Fig. 3.12 *AAL-5 'Level Mixing' with the ATM-PDU PTI*

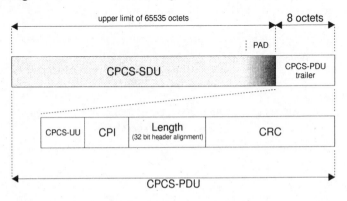

Fig. 3.13 *CPCS-PDU Frame Structure for AAL-5*

Convergence Sublayer Frame coding

Only the CPCS-SDU and a trailer exist within the CPCS-PDU; see Figure 3.13. The trailer consists of 1-octet CPCS User-to-User, 1-octet Common Part Indicator, 2-octet Length and 4-octet Cyclic Redundancy Check (CRC).

The CPI field is currently undefined and only used to align the CPCS-PDU trailer to 64 bits. The user-to-user information in the CPCS-UU parameter of the CPCS-UNITDATA-invoke and CPCS-UNITDATA-signal is carried to its peer entity by means of the CPCS-UU field.

Error detection of the entire SAR-PDU is covered by a 32-bit CRC. The CRC is the remainder of the product of the entire CPCS-PDU (except the CRC) and x^{32}, divided by the generator polynomial

$$x^{32} + x^{26} + x^{23} + x^{22} + x^{16} + x^{12} + x^{11} + x^{10} + x^8 + x^7 + x^5 + x^4 + x^2 + x + 1.$$

The CRC will detect bit errors, misinserted, misdirected or missing SAR-PDUs, therefore no further checking is required. However, the maximum probability of the CRC failing to detect any number of bit errors is 2^{-32}. The Length field decreases this probability by a further factor of 2^{-16} (approximately 10^{-5}). The Length Indicator is a binary representation of the number of octets in the CPCS-SDU; this limits AAL-5 to an absolute maximum CPCS-PDU size of 65 535 octets.

To fill all SAR-PDUs to the 48-octet boundary, the PAD field performs 48-octet alignment, that is, PAD is between 0 and 47 octets. A 47-octet overhead might seem significant but the ATM cells would otherwise carry blank information and the 47-octet PAD field therefore induces no additional overhead on the transmission link, though buffering at the destination is slightly increased.

3.5   AAL-3/4 and Class D (Connectionless)

The ability of the AAL to support connectionless services over ATM requires a high level of AAL interaction. I.211 (reference [22]) divides the connectionless service into two types:

- *Indirectly via a B-ISDN connection-oriented service.* This service provides a connection between end-points. It is a function of the higher layers, above B-ISDN, to provide the connectionless protocols.
- *Directly via a B-ISDN connection-oriented service.* The Connectionless Service Function (CLSF) is handled by an additional layer above the *CLSF* adaptation layer. For the UNI, the CLSF is provided by Connectionless Network Access Protocol (CLNAP) and at the NNI the Connectionless *CLNAP* Network Interface Protocol (CLNIP) is implemented. Both are recom- *CLNIP* mended in I.364 (reference [23]) although the CLNIP requires further study in the standards bodies.

Two reference points have been recommended: for indirect CLSF, the M or P reference point is used and for direct CLSF the P reference point is used. See Figure 3.14.

3.5.1   User interface

Two primitives exist at the CLNAP SAP called *CLNAP-UNITDATA.request* and *CLNAP-UNITDATA.indication*. All the parameters are transferred to the CLNAP frame; see subsection 3.5.2 for a description of their function. As mentioned above, the CLNIP user interface requires further study.

The CLNAP primitives, all of which are mandatory, are as follows:

- CLNAP-UNITDATA.indication which contains:
  - Source address
  - Destination address
  - Data
  - QoS

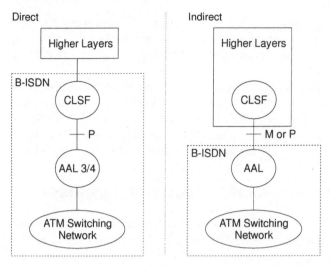

Fig. 3.14 *CLSF provided directly and indirectly*

- CLNAP-UNITDATA.request which contains:
  - Source address
  - Destination address
  - Data
  - QoS

3.5.2 CLNAP Frame structure

The CLNAP Frame structure is shown in Figure 3.15. The 64-bit source and destination address fields are divided into a 4-bit *address-type* subfield and a 60-bit *address* subfield. This allows for differentiation between a 60-bit publicly administered individual or group address. The structure of the 60-bit address field is in accordance with E.164 (reference [24]). Further addressing at the destination is provided by the 6-bit Higher-Layer Protocol Identifier (HLPI) field. This indicates the destination higher layer entity CLNAP-SDU.

HLPI

Additional information specific to the service/connection can be transported across the network in the variable length Header Extension field. This field is divided into a 1-octet Element Length, 1-octet Element Type and variable length Element Payload subfields. The Element Payload information is determined by the Element Type although the use of both fields requires further study. To calculate the length of the Header Extension field the Element Length subfield contains the length of the entire Header Extension field in octets.

Management information for Quality of Service parameters is transferred across the network in the 4-bit QoS field. QoS can be altered by implementing a frame CRC. If the CRC indication bit (C) is equal to 1, this indicates the

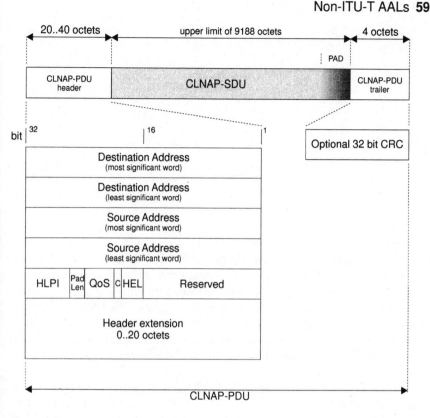

Fig. 3.15 *CLNAP frame structure*

presence of the optional 32-bit CRC in the trailer. The CRC is the remainder of the product of the entire CPCS-PDU (except the CRC) and x^{32}, divided by the generator polynomial

$$x^{32} + x^{26} + x^{23} + x^{22} + x^{16} + x^{12} + x^{11} + x^{10} + x^8 + x^7 + x^5 + x^4 + x^2 + x + 1.$$

The PAD field keeps the entire frame aligned to a 32-bit boundary and therefore ranges in length from 0 to 3 octets depending on the CPCS-SDU length.

3.6 Non-ITU-T AALs

Various other AALs have been suggested by academia and industry. These AALs improve on the AALs currently defined and their benefits have been shown by simulations or mathematical models. The suggested AALs generally improve the service interaction performance in various areas although most pay penalties, particularly those systems implementing cell recovery techniques. Reference [2] analyses AAL-5 and recommends a 34-bit CRC and a 4-state HEC

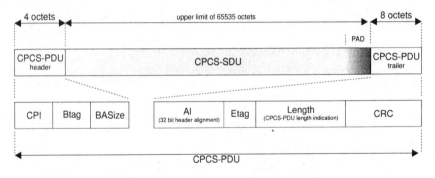

Fig. 3.16 *Modified AAL-3/4 CPCS PDU structure*

error correction/detection scheme for the ATM layer. Reference [25], amongst others, studies the possibility of implementing cell recovery techniques.

BLINKBLT   A particularly interesting AAL was developed by Bolt Beranek & Newman Inc. called BLINKBLT (Broadband Link-Layer Block Transfer). Its current status is defined in ANSI contribution T1S1.5/90-009. The protocol was intended for services requiring assured mode data transfer and implements, unlike AAL-3/4 and AAL-5, individual cell (SAR-PDU) Automatic Repeat Request

ARQ   (ARQ).

The protocol is made compatible with AAL-3/4 by coding the Segment Type field of the SAR-PDU to a value greater than 48. The BLINKBLT-PDU combines the Sequence Number and Multiplex Identification field to form a 14-bit Sequence Number field.

The CPCS-PDU has an additional 32-bit CRC field that detects all otherwise undetected errors from the SAR sublayer, such as sequencing errors, bit errors and cell loss. SAR-PDUs are acknowledged using a bitmap status of the received SAR-PDUs.

BLINKBLT has never been implemented since it would only be efficient over AAL-3/4 if connections had a high cell loss. A high cell loss would only be applicable to unspecified bit rate service provision. Available Bit Rate service provision has since been recommended which reduces cell loss.

Similar to BLINKBLT, a modified AAL-3/4 has been developed by the present authors, with improved characteristics; see reference [26]. This modified AAL-3/4 includes a 34-bit CRC to the CPCS-PDU, shown in Figure 3.16. The error detection coverage of this 34-bit CRC relieves the SAR-PDU's CRC of its error detection requirements and allows a smaller error detecting/correcting code, such as the 9-bit Hamming code shown in the SAR-CPCS PDU in Figure 3.17. The extra information bit has been shown to be useful in increasing the error detecting ability of the SN field. It has been shown that the modified AAL-3/4 can offer significant advantages through its error correcting ability.

A further extension, made by the present authors, to the modified AAL-3/4

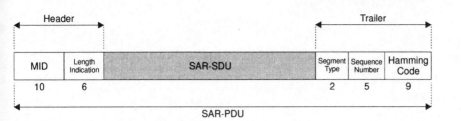

Fig. 3.17 *Modified AAL-3/4 SAR PDU structure*

includes cell loss recovery techniques. The recovery technique is based on a unique Automatic Error Code Repeat Request (AERQ) system, whereby the error code is only transmitted when an erroneous cell is detected. The error code is a (128, 124) Reed–Solomon Code, see Figure 3.18, similar to the code used in AAL-1. For accepted network error rates these inclusions can improve the worst case frame retransmission rate from 9 minutes to 256 years (see reference [27]).

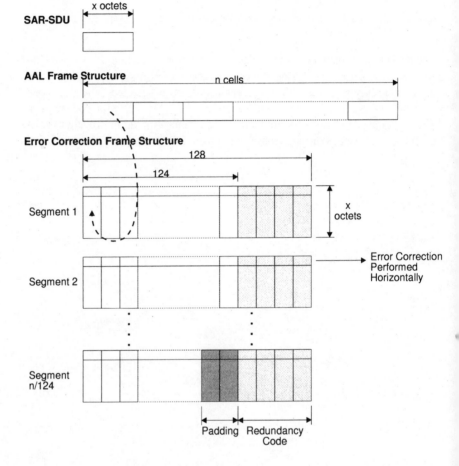

Fig. 3.18 *AAL and Reed–Solomon frame structure*

4
ATM SIGNALLING

4.1 Introduction

It is not the intention of this chapter to detail the signalling protocols because
such description would be out-of-date before the authors had finished writing
and there are a few 'monster' books published detailing every signalling issue.
The idea here is to convey enough information for the reader to gain a basic understanding of the requirements, behaviour and complexity of signalling
protocols within B-ISDN.

For a very basic description of signalling, see the entries in the glossary
for signalling and switched connection. Primarily, signalling is about setting-up
and clearing temporary connections through the network, permanent connections
being established through management functions. These temporary connections
are called Switched Virtual Channels (SVCs), and differ from Permanent Virtual *SVC*
Channels (PVCs) in that some form of signalling is required to establish and *PVC*
remove them.

An SVC or PVC need not connect a single source to a single destination but
may instead connect a single source to multiple destinations or multiple sources
to multiple destinations. These different topologies are known as connection
types and are more formally called point-to-point (pt–pt), point-to-multipoint *pt–pt*
(pt–mpt) and multipoint-to-multipoint (mpt–mpt). *pt–mpt*

For B-ISDN, defining the signalling protocols is probably the most complex *mpt–mpt*
process the standards organisations have to achieve. Nevertheless it has been
possible to constrain the problem and implement only a small set of the signalling protocols – generally the most basic ones – even at the expense of
limiting the functionality of the network. For example, B-ISDN is intended
to support multipoint-to-multipoint connections but it is feasible to operate a
point-to-point network while the multipoint-to-multipoint signalling protocols
are being defined.

The signalling protocols have already taken several years of development and
have been updated through the emergence of several standards. As with other
aspects of B-ISDN, the ITU-T and ATM Forum are responsible for the development of the signalling procedures and protocols. At the User–Network Interface

(UNI), the basic call connection control standards have been developed from the ITU-T's Q.931 Narrowband ISDN signalling protocols (reference [28]). The more comprehensive set of standards, incorporating B-ISDN, is called Q.2931 (reference [67]). Point-to-multipoint connection control is defined in Q.2971.

B-ISUP
SS7

At the Network–Network Interface (NNI) the B-ISDN protocol is defined in the Broadband ISDN User Part (B-ISUP), an addition to the Narrowband ISDN Signalling System Number 7 (SS7).

The ATM Forum has made major contributions to the development of signalling. It first introduced SVCs in its UNI 3.0 standard and these were expanded in UNI 3.1 (reference [11]) and further enhanced in UNI 4.0 – this last being a standard specifically to do with signalling. Although UNI 4.0 Signalling refers to Q.2931, there are still gaps in interoperability between the two standards. These gaps are of immediate concern to the ITU-T and ATM Forum.

PNNI

The ATM Forum in its Private Network–Network Interface (PNNI) standard is also developing NNI signalling.

The following release format shows how the ITU-T intends gradually to implement the signalling procedures.

- Release 1 supports the signalling for constant bit-rate, connection-oriented services with end-to-end timing (i.e. circuit emulation). This may be implemented over symmetrical and asymmetrical point-to-point connections. The signalling protocols inter-work with Basic Rate $(2B + D)$ ISDN and Primary Rate services and provide for meta-signalling, a technique for establishing signalling channels in Virtual Paths.

- Release 2 enhances release 1 by incorporating Variable Bit Rate connection-oriented services with the Quality of Service indicated by the user. The connection may also be point-to-multipoint, and negotiation and re-negotiation of bandwidth is possible.

- Release 3 is the final implementation that will support multimedia and distributive services with Quality of Service negotiation. The connection may also be broadcast.

4.2 The Signalling AAL

Before any signalling protocols can be used, the method of transporting the signalling information between the network switches must be defined. Intuitively, this seems to be a waste of effort since the ATM adaptation layers have already been defined to provide end-point to end-point communication. It is, however, of utmost importance that the signalling information sent between two points be accurate, and the currently defined AALs do not provide this level of support. Therefore a Signalling ATM Adaptation Layer (S-AAL) has been defined.

S-AAL

The S-AAL has the same function as the High-Level Data Link Control Protocol (HDLC) based Link Access Protocol for D channel (LAP D) used to carry signalling information in Narrowband ISDN.

HDLC
LAP D

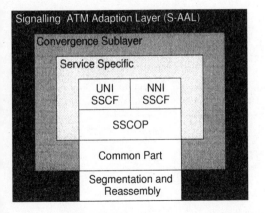

Fig. 4.1 *Signalling-AAL layered architecture*

The S-AAL is intended for both UNI and NNI signalling, although slightly different formats apply. The structure of the S-AAL is summarised in Q.2100 (reference [29]). Signalling information within an ATM network is sent along Virtual Channels within Virtual Paths in ATM cells just like any other ATM traffic. The identities of the channels and paths (VCIs and VPIs) are *well known* (i.e. predefined).

The protocol reference model of the signalling layers at the UNI is shown in Figure 4.2. The S-AAL is subdivided into 3 sublayers similar to AAL-3/4 and AAL-5 (see Figure 3.8).

The Common Part (CP) and Segmentation And Reassembly (SAR) sublay- *CP* ers are implemented using the AAL-5 protocol, although the AAL-3/4 protocol *SAR* would also be suitable. The Service-Specific Part (SSP) is further divided into *SSP* 2 sublayers as shown in Figure 4.1. The Service-Specific Connection-Oriented Protocol (SSCOP), see Q.2110 (reference [30]), is common to UNIs and NNIs *SSCOP* and the Service Specific Co-ordination Function (SSCF), see Q.2130 (refer- *SSCF* ence [31]), is UNI or NNI specific.

4.3 Signalling at the UNI

UNI signalling occurs at the UNI interfaces described in Table 1.3 to allow calls to be established and cleared in a controlled manner. In order for this signalling to operate, two requirements must be met: there must be a clearly defined protocol over well-known VCI/VPIs and there must be a means of uniquely identifying each terminal attached to the network (the equivalent of a telephone number).

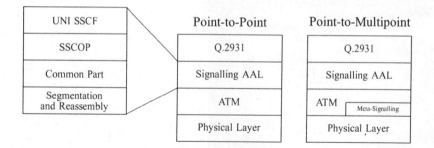

Fig. 4.2 *UNI Signalling Reference Model*

4.3.1   UNI signalling protocol

The UNI signalling protocols are based on a simplified version of the basic and primary rate ISDN signalling protocols (see Q.931 – reference [28]) and have been recommended in Q.2931 (reference [67]) for point-to-point connections. Point-to-multipoint connections use a *meta-signalling* protocol to manage the multiple signalling relations.

As an example of the UNI signalling interface, Figure 4.3 shows the messages exchanged between a calling party, the network and the called party, in order to establish a call. Note that this figure does not show everything that may go wrong during the set-up: call being rejected, timers expiring, etc. In this diagram the network is shown as a single entity. In practice, of course, the network will probably consist of many switches and the call request has to be routed through the network from source to destination: see subsection 4.4.2.

The initial *SETUP* message contains such information as the address of the destination, the required Quality of Service parameters, message type and details of the higher-level protocols being carried (so that these may be terminated at the other end).

Once the *SETUP* message has been sent, a 4-second timeout (T303) is started to protect the calling party from network failure. On receipt of the *SETUP* message, the network may return a *CALL PROCEEDING* to stop T303 and start a longer timeout (T310) if it is not going to be able to establish the call within 4 s, and the handshaking continues, as shown in Figure 4.3.

Note that the call set-up messages are, as far as the calling and called parties are concerned, symmetrical: for a call to be established a network is not required.

4.3.2   Addressing

Well-structured address formats are very important to large networks. Local area networks typically have a simple, flat, unstructured addressing scheme (Media

MAC   Access Control (MAC) addresses) which works well in an environment with

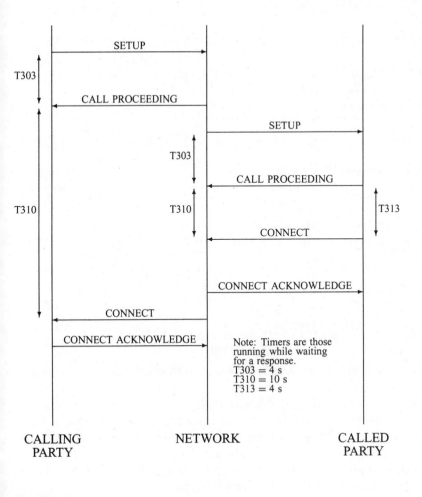

Fig. 4.3 *UNI call set-up messages*

only a few connected hosts but not where there is a large number of addressed devices. In a larger network some form of hierarchical addressing is required.

Within private ATM networks, the addresses of the various terminals are modelled on the format of the address of an ISO OSI Network Service Access Point (NSAP) (see ISO-8348 and X.213 – references [32] and [33]). For a tutorial on the complex matter of address assignment, see RFC1237 – reference [34].   *NSAP*

Within the NSAP address format, three types of hierarchical address are used, distinguished by an Authority and Format Identifier (AFI) as the first octet of the address. The three forms of the address are shown in Figure 4.4 where the abbreviations used have the following meanings: *AFI*

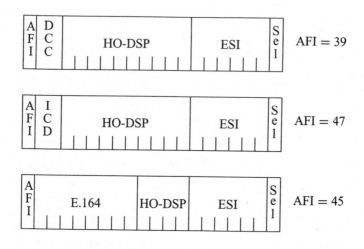

Fig. 4.4 *UNI address formats*

DCC	**DCC.** Data Country Code. A 2-octet field which specifies the country, as defined in ISO-3166 (reference [35]).
ICD	**ICD.** International Code Designator. Another national identifier, maintained by the British Standards Institute.
E.164	**E.164.** This cites reference [24] which specifies Integrated Services Digital Network numbers (i.e. ISDN numbers, including telephone numbers). These numbers can be up to 15 digits in length and are stored in Binary-Coded Decimal (BCD) format.
HO-DSP	**HO-DSP.** High-Order Domain-Specific Part. This field is used, by the authorities issuing the AFI, to specify a subdomain.
ESI	**ESI.** End-System Identifier. This 6-octet field identifies the particular terminal within the specified network. It is recommended that the 48-bit MAC address of the terminal be used to simplify the support for LAN equipment, where the MAC address is built into the hardware.
Sel	**Sel.** Selector. This is a 1-octet field, not needed for the ATM routing, which may be used by the end-systems themselves.

Within public networks, terminals may be identified either by a structured number, as in a private network, or by a pure E.164 (reference [24]) number.

To avoid having to register every address, the ATM Forum has defined a protocol which allows the ATM terminal equipment to inform its ATM switch of its address dynamically across the UNI. Basically the terminal's MAC address is sent from the terminal to the switch and the switch responds with the remaining bits of the full address. This technique is known as *address registration*.

4.3.3 Meta-signalling

The idea of the meta-signalling protocol is to allow one signalling channel per VP. Unlike Q.2931 and B-ISUP, the meta-signalling protocol (see Q.2120 – reference [36]) has only one signalling protocol format. The elements that are not needed in specific messages are ignored and normally filled with NULL characters. The general format is shown below in Figure 4.5. The Protocol Identifier is used to determine whether the message is for the meta-signalling protocol or for other signalling protocols using the meta-signalling channel. The style of message is represented in the Message Type field. There are currently six types of message used within the meta-signalling protocol:

- Assign Request: The initial message generated by the user for the assignment procedure.
- Assigned: An acknowledgement and acceptance of an Assigned Request message. A point-to-point Signalling Virtual Channel and a Broadcast Signalling Virtual Channel (BSVC) are associated with the connection.
- Denied: An acknowledgement and rejection of an Assigned Request message.
- Check Request: The initial message generated by the network plane management for the check procedure. Only one check procedure is allowed at any one time.
- Check Response: An acknowledgement of a Check Request.
- Removed: Two Removed messages are sent, separated by a random time interval. The message is initiated by the user or network.

To differentiate between assignment procedures occurring simultaneously, the reference identifier is given a randomly generated value. The response to each assignment is then identified by the associated identifier.

The style of signalling on the channel is identified by the Signalling Virtual Channel Identifier A and B. The three possible signalling connections are point-to-point, broadcast or global signalling (PSVCI, BSVCI, GSVCI respectively). *PSVCI*

Point-to-multipoint and point-to point configurations are distinguished by the *BSVCI* Signalling configuration field. Currently the point-to-point Cell Rate field can *GSVCI* only indicate a peak cell rate allocation for the signalling channel. This should change in future releases to accommodate a request/allocated architecture and to allow the introduction of this, the Protocol Version field is used as a differentiator.

To provide more information about any errors or actions the Cause field is used. The user identifies the level of service required by means of the Service Profile Identifier (SPID). This points to a set of network-maintained service *SPID* profiles.

The complete frame is covered using a 32-bit CRC error detection code. The CRC is the remainder of the product of the entire ATM-SDU (except the CRC)

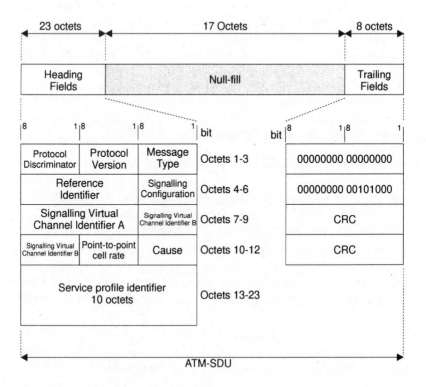

Fig. 4.5 *Meta-signalling cell format*

and x^{32}, divided by the generator polynomial

$$x^{32} + x^{26} + x^{23} + x^{22} + x^{16} + x^{12} + x^{11} + x^{10} + x^8 + x^7 + x^5 + x^4 + x^2 + x + 1$$

4.4 Signalling at the NNI

4.4.1 NNI signalling

Two types of NNI signalling are being defined by the ATM Forum and ITU: a private NNI (PNNI) for use between switches in a private network and a public NNI for use between public networks (using E.164 addresses: see sub-section 4.3.2).

Public NNI protocols, based around E.164 addressing, are implemented by the B-ISUP – see Q.700 (reference [37]).

The definition of the PNNI consists of two parts:

- a protocol part, an extension of the UNI protocol, used across the network to indicate the user's request to establish or clear a call
- a routing part, which determines how the user's connection is routed across the network (see the following section).

4.4.2 Routing

Because ATM is a connection-oriented protocol it needs to find a path through the network when a call is established (in contrast to connectionless protocols where each transmitted packet is routed separately).

At call establishment, a message has to be sent through the network from the caller's switch to the called party's switch reserving the necessary (and guaranteed) bandwidth for the call. Although ATM is connection-oriented, connection-establishment messages have to be sent in a connectionless mode as no connection exists at that time. The search for the 'best' route (even if the concept of 'best' could be defined) is very difficult because of the Quality of Service parameters (cell loss ratios, cell delays, etc.) which must be met at every switch and on every link of the path.

The first level of call establishment is Connection Admission Control (CAC). *CAC* The principle is for the originating user's switch to determine, *before* accepting the call, that a route can be established through the network, meeting the required QoS parameters (see Figure 4.3, where this happens between the arrival of the *SETUP* message at the network and the transmission of the *CONNECT* back).

Since there is no single system in the network with a knowledge of all of the switches' and links' loads, this information has to be shared between all switches: the information being exchanged in PNNI Topology State Packets (PTSPs). Given a set of PTSPs, received from all the other switches in the *PTSP* network by means of a network-flooding technique, a switch can make a stab at finding a path through the network which will probably meet the specified Quality of Service parameters of the call. The technique whereby the source switch determines the whole route (rather than simply firing a packet into the network to be routed hop-by-hop) is known as Source Routing. *Source*

Any route found like this must at best be a guess since information received *Routing* through PTSPs will typically be somewhat out of date: particularly in a network with a high rate of call set-ups. Once a route has been determined, an attempt is made to establish the call by sending a message along the path, reserving the necessary bandwidth *en route*. If a particular switch is unable to guarantee the Quality of Service for the call, then it rejects the set-up and the call itself must fail or another route must be attempted.

The area of Connection Admission Control is one where much research is taking place and one where different vendors will be able to differentiate their products.

5
OPERATION AND MAINTENANCE (OAM)

5.1 Basic principles

The ITU-T has defined a set of Operation and Maintenance (OAM) functions for B-ISDN and published the intent that:

> A B-ISDN will contain intelligent capabilities for the purpose of providing advanced service characteristics, supporting powerful operation and maintenance (OAM) tools, network control and management.

This is an ambitious task not yet completed. In this chapter it will be addressed in two parts: where measurements are taken to assess the network's performance and where the resulting information is correlated and presented to a network operator. A useful introduction to the techniques involved in the latter is the collection of papers gathered together in reference [38].

The underlying principle of OAM at both levels is the concept of a *Network Element* (NE). Each object in the network which is to be monitored or controlled *NE* is a Network Element and one of the major problems facing the designers of OAM systems is how to handle the variety of Network Elements within any one network: from small, dumb repeaters to highly sophisticated voice and data switches. The problem is accentuated by the heterogeneous collection of equipment from different manufacturers that normally comprises a network. The need for some form of standardisation is clearly essential to enable networks to be managed.

5.2 The OAM flows

For information gathering, the standards assume that the following OAM functions will need to be performed (see I.610 – reference [12]):

- Performance Monitoring: monitoring network elements to gather maintenance information. This informs the operator of overloaded or under-utilised parts of the network.

- Failure Detection: generating alarms on malfunction or predicted malfunction of equipment.

- System Protection: minimising the effect of a malfunction by automatically bypassing a piece of failed equipment.

- Failure or Performance Information: communicating failure information to higher management planes and responding to requests for status reports.

- Fault Localisation: tracing the cause of a fault by means of internal or external test systems. This is of particular value to a network operator as many faults produce multiple alarm messages which can mask the actual cause of the fault – dispatching a repair technician to the wrong site can be very expensive.

F1–F5   In order to subdivide the work of collecting this information, five hierarchical levels have been identified: unoriginally named F1 to F5.

F1, F2 and F3 apply to sublayers of the physical layer and F4 and F5 apply to the ATM Virtual Path and Virtual Channel layers respectively.

These flows are illustrated in Figure 5.1 (taken from I.610 – reference [12]) where a Virtual Channel passes (from left to right) through some form of signal regenerator, through a Virtual Path Switch and through a Virtual Channel Switch before being terminated. For an illustration of Virtual Channel and Virtual Path Switching, see Figure 1.4.

The functions of the flows F1 to F5 are, in summary:

F1. This flow covers a single regenerator section of the transmission path. It detects loss of signal and loss of frame alignment from B-NT1 or B-NT2 (see Figure 1.8).

F2. In non-ATM systems, F2 also covers a single regenerator section of the transmission path but at a higher level than F1: rather than simple failures, F2 detects unacceptable error performance. In ATM systems the F2 flow is not used, its function being absorbed into F3.

TP   **F3.** This flow covers an entire digital transmission path (TP) (which may consist of several digital sections end-to-end), covering cell rate decoupling, cell delineation, Customer Network status monitoring and loss of AU Pointer alignment. Idle cell errors and loss of cell synchronisation are also detected at this level.

F4. This operates at the Virtual Path level: monitoring for path availability and detecting unavailable paths and degradation in system performance.

F5. This provides the end-to-end monitoring of the Virtual Channel, checking for channel availability and monitoring performance to detect unavailable paths and system degradation.

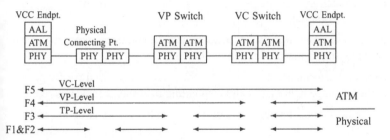

Fig. 5.1 *OAM flows F1–F5*

5.3 Physical Layer flows

The Physical Layer flows F1, F2 and F3, by their passage, indicate that the physical link is operational in the areas of signal detection, frame alignment, section error monitoring, cell rate decoupling, cell delineation, customer network status monitoring and AU-4 pointer operation.

Additionally, Network Management systems may use Physical Layer flows to provide performance monitoring and reporting and localisation of failed equipment. This is dealt with in more detail in section 5.5 below.

None of the cells which form the F1 and F2/F3 flows are passed to the ATM layer.

When ATM is used across an SDH-based (or SONET-based) physical layer, OAM flow F1 is carried in the SDH Section Overhead (SOH) (see Figure 2.7) and flow F3 is carried in the SDH Path Overhead (POH) as defined in G.707 (reference [39]).

5.4 ATM Layer flows

Flows F4 and F5 are subdivided into End-to-End and Segment flows.

End-to-End F4 flows operate between VPC endpoints, whereas Segment F4 flows operate across one or more concatenated VPC links.

In a manner analogous to F4, End-to-End F5 flows operate between VCC endpoints, whereas Segment F5 flows are implemented across one or more concatenated VCC links. Both F5 flows are identified by a standardised Payload Type Identifier (PTI), see section 2.7 and I.361 (reference [15]).

All types of F4 and F5 flow are further identified by their appearance on different but standard VCIs; see Table 2.3 in section 2.7 and I.361 (reference [15]).

When an F4 or F5 flow passes through intermediate switches, those switches may monitor the flow (and may even inject additional cells) but not terminate the flow.

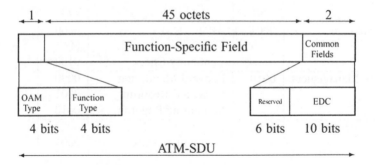

Fig. 5.2 *OAM common cell format*

For F4 and F5 flows, two messages are defined:

AIS

- An Alarm Indication Signal (AIS), either at the VPC or VCC level, which is sent downstream by a device that detects an incoming physical-level failure. These failures result in either a VP-AIS or VC-AIS.
- A Far-End Receive Failure Signal (FERF) which is sent to the remote end of the VP or VC by a device that has received an AIS. As with AISs, FERFs can be categorised as VC-FERFs or VP-FERFs.

One of the factors considered by a switch when accepting a new call connection is the availability of bandwidth not only for the customer's traffic but also for OAM cells.

5.4.1 The OAM cell

There is a common cell format for all ATM layer OAM cells: see Figure 5.2. The cell is identified by its OAM type, for example fault management or performance management, and its function type as listed in Table 5.1.

EDC

Error detection of the entire ATM-SDU is covered by the Error Detection Code (EDC), which is a 10-bit CRC. The CRC is the remainder of the generator polynomial $x^{10} + x^9 + x^5 + x^4 + x + 1$ divided by the product of the entire ATM-SDU (except the CRC), and x^{10}.

5.4.2 Alarms

A VP-AIS or VC-AIS is generated as soon as possible after a failure is detected and, to avoid rapid generation and clearing of the alarm, is held for a nominal period of 1 s even if the failure disappears almost immediately. In the example shown in Figure 5.3, it is assumed that a VCC or VPC is established between user A and user B and that the link from node 1 to node 2 fails. The Physical

OAM type	Binary code	Function type	Binary code
Fault Management	0001	AIS	0000
		FERF	0001
		Continuity Check	0100
Performance Management	0010	Forward Monitoring	0000
		Backward Reporting	0001
		Monitoring/Reporting	0010
Activation/Deactivation	1000	Performance Monitoring	0000
		Continuity Check	0001

Table 5.1 *OAM type and function type codes*

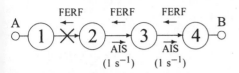

Fig. 5.3 *AIS/FERF sequence*

Layer of node 2 will detect this and node 2 will start to transmit AIS messages downstream (i.e. to node 4, their being monitored *en route* by node 3) at the rate of 1 per second.

At the receiver, node 4, a VP-AIS or VC-AIS status is declared as soon as a cell is received, and cleared either after 3 seconds without receipt of AIS cells, or upon receipt of one valid user or continuity check cell. The format of the AIS function-specific field can be seen in Figure 5.4 and the cell coding structure in Table 5.1. The details of the function-specific fields have not been specified and all fields default to 0x6A.

When a VP-AIS state or a VPC failure is detected, a VP-Far-End Receive Failure (VP-FERF) is generated and sent to the far-end, travelling from node 4 *FERF* to node 1 (if it is reachable) as shown in Figure 5.3. Similarly a VC-FERF is generated when a VC-AIS state is declared or a VCC failure is detected.

FERFs are transmitted periodically, nominally once per second, until the failure condition disappears.

A VP-FERF or VC-FERF status is declared when one FERF cell is received, and cleared after 3 s without receipt of these cells. Thus, in the example of Figure 5.3, if the failure affects only uni-directional traffic from node 1 to node 2, node 1 will receive the FERF and declare a VC-FERF or VP-FERF status. The format of the FERF function-specific field can be seen in Figure 5.4 and the cell coding structure in Table 5.1.

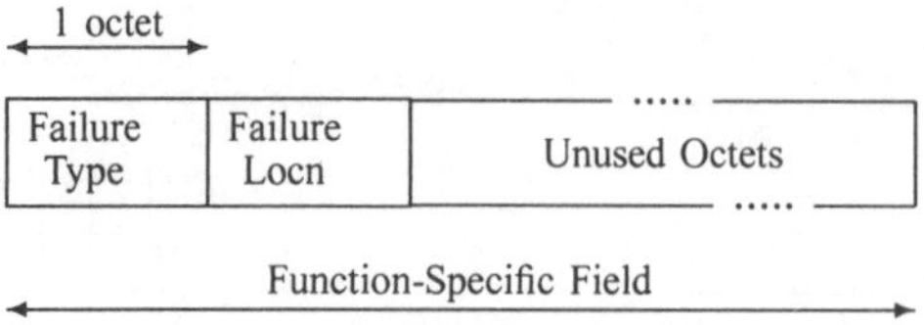

|← 1 octet →|

Failure Type	Failure Locn	..... Unused Octets .....

←————————— Function-Specific Field —————————→

Fig. 5.4 *AIS/FERF Function-Specific Cell Field*

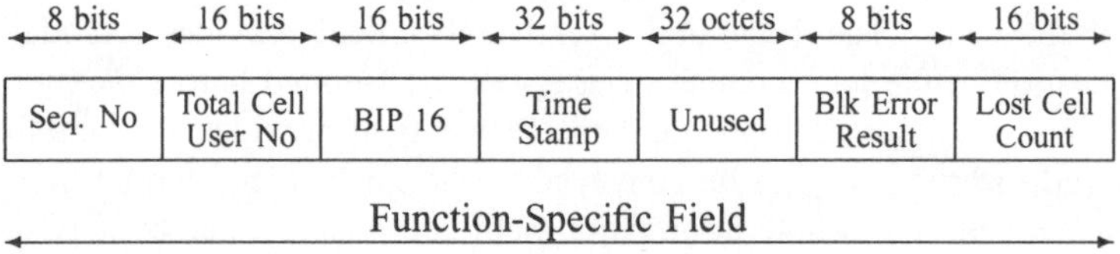

|← 8 bits →|← 16 bits →|← 16 bits →|← 32 bits →|← 32 octets →|← 8 bits →|← 16 bits →|

Seq. No	Total Cell User No	BIP 16	Time Stamp	Unused	Blk Error Result	Lost Cell Count

←————————————— Function-Specific Field —————————————→

Fig. 5.5 *Performance Monitoring Function-Specific Field*

The details of the function-specific fields have not been specified and all fields default to 0x6A.

5.4.3 Performance Monitoring

The primary objectives of Performance Monitoring (PM) are to detect corrupt, lost or misinserted cells. This is achieved by inserting monitoring cells into a VPC or VCC.

Performance Management is activated and deactivated by the Telecommunication Management Network (TMN) (see section 5.5). On activation, one VPC or VCC end point sends an ACTIVATE PM cell to the other end to request monitoring. The receiving end can accept or decline the invitation by returning a PM ACTIVATION CONFIRMED or a PM ACTIVATION REQUEST DENIED cell respectively.

Once the end points have agreed to monitor performance, a performance monitoring cell, see Table 5.1, is inserted into the VPC or VCC end-to-end or over a segment depending on the flow being monitored. PM cells are inserted into the data stream for the VPC or VCC every N cells where $N \in \{128, 256, 512, 1024\}$. The actual number of cells between insertions can vary up to 50% for end-to-end performance monitoring. For end-to-end monitoring, the monitoring cell must be inserted into the user cell stream within $N/2$ cells after the insertion request; this is known as a forced insertion.

The Performance Monitoring Function-Specific Field, see Figure 5.5, contains a Monitoring Cell Sequence Number which, modulo 236, provides a sequence

6 bits	2 bits	8 bits	4 bits	4 bits	42 Octets
Msg. ID	Directions of Action	PM Block Size A-B	Correlation Tag	PM Block Size B-A	Unused

Function-Specific Field

Fig. 5.6 *Activation/Deactivation Function-Specific Field*

number for the cells. The Total Cell User Number field provides a modulo 65 536 indication of the total number of user cells transmitted before the monitoring cell was inserted. An optional Time Stamp field provides a means for the OAM cell insertion time to be conveyed to the far end, the format being currently undefined. The whole information field is error protected by a 16-bit Bit Interleaved Parity (BIP) field.

For backward reporting, the Block Error Result informs the originating element of any errors detected by the BIP field. The detection of loss or misinserted cells is conveyed by the Lost/Misinserted Cell Count Field.

5.4.4   Continuity Checks

This area is still under study but its purpose is clear: to ensure that periods when no cells are transmitted are due to the users not transmitting rather than connection failure. Following a handshake between the two ends, special Continuity Check Cells, see Table 5.1, are artificially generated after a period of silence on a VCC. The endpoint can then raise VC-FERF if it does not receive any cells (user or Continuity Check Cells) within a certain time.

5.4.5   Activation and Deactivation

Activation/Deactivation is used for establishing Performance Monitoring (see subsection 5.4.3) or Continuity Checking (see subsection 5.4.4). The cell format consists of six fields as shown in Figure 5.6. The Message ID specifies the message's function; see Table 5.3.

The direction of the cell (A → B or B → A) is held in the Directions of Actions field which holds four values; see Table 5.4.

Command and response messages can be matched as they have the same value in their Correlation Tag fields. The size of the block used for performance monitoring is given for each direction in the PM Block Size A-B and PM Block Size B-A fields, as shown in Table 5.2. All unspecified octets have the value 0x6A.

	128	256	512	1024
Example/Bit Number	bit 4	bit 3	bit 2	bit 1
512 only	0	0	1	0
256 only	0	1	0	0
128 and 1024 only	1	0	0	1
256, 512 and 1024 only	0	1	1	1

Table 5.2 *PM Block Size field coding*

Message	Binary Coding
Activate	000001
Activation Confirmed	000010
Activation Request Denied	000011
Deactivate	000101
Deactivation Confirmed	000110
Deactivation Request Denied	000111

Table 5.3 *Message ID field coding*

Direction	Binary Coding
A → B	10
B → A	01
Bidirectional	11
Default (N/A)	00

Table 5.4 *Direction of Actions field*

5.5 Telecommunications Management Network

5.5.1 Introduction

In the previous sections we examined some of the mechanisms for gathering information about a network. Once gathered, it must be correlated and displayed in a form suitable for a human operator. The set of standard protocols, interface points and architecture needed to achieve this is called a Telecommunications *TMN* Management Network (TMN).

The functions which network owners need to be able to control are often *FCAPS* abbreviated to *FCAPS* (pronounced 'ef-caps'). The acronym comprises the initial letters of:

- Fault Management: display of alarms arising in the network, fault localisation and testing.
- Configuration Management: ability to provision elements of the network, perform backups, load new software, take inventories, etc.

- Accounting Management: collection of billing data, etc.

- Performance Management: collection of performance data, analysing that data to spot trends, etc.

- Security Management: prevention of unauthorised access to network elements (including the elements of the TMN itself: workstations, databases, etc.).

The idea of a unified Network Management architecture has been an elusive dream for many years; many telecommunications companies have produced such products which have eventually collapsed, normally under their own weight. As stated above, the problem is very difficult: extracting the FCAPS information from a heterogeneous set of thousands of network elements, manufactured by different companies at different times and widely scattered geographically, and displaying that information homogeneously for a human operator.

Within the ITU, the standardisation of TMN began in 1985 and still continues. As with most concepts in telecommunications, layering is once again a feature of the TMN. Management of a network is viewed at several different levels, corresponding to the following layers:

- Business Management Layer. This layer provides network owners with network-wide information to allow them to make strategic decisions about the growth of the network.

- Service Management Layer. This layer supports the FCAPS functions for end-to-end customer services: how often they failed, which ones should be set up, what a customer should be billed, etc. This layer contains no information about physical connections or switches; it deals with services as perceived by a customer.

- Network Management Layer. This layer is responsible for collecting and collating the data from individual Network Elements to provide a network-wide view for the operator.

- Network Element Management Layer. This layer functions as the Network Management Layer but typically only for a small subset of the Network Elements in the network: perhaps those in a particular location or those from a particular manufacturer.

- Network Element Layer. This layer has no knowledge of the wider picture, it deals with FCAPS for one Network Element. The information that it supplies is collected by the Network Management Layer to provide an operator with a coherent view of the network.

Note that the overall management of a network is often referred to as *Network Management* but, strictly speaking, that term refers only to one of the five layers listed above.

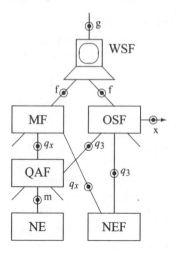

Fig. 5.7 *Telecommunication Management Network architecture*

5.5.2 Architecture

The TMN specifies a set of functional groups and reference points. This allows a generic management architecture to interface with different real network architectures. The basic structure of a TMN is shown in Figure 5.7 where standardised reference points are shown by a dot with concentric circle and labelled with a lower-case letter. Most of these reference points have standard interfaces associated with them (represented by the corresponding upper-case letter) which are used when the two ends of the interface are implemented in separate devices. The exceptions are the 'g' reference point, which lies between a workstation and a human operator, and the 'm' reference point which lies between the TMN and non-TMN devices. The 'G' interface is left to the imagination of individual manufacturers and the 'M' interface has to be tuned to the specific Management interface offered by the non-TMN device.

The functions of each group are summarised here, but for a full description see M.3010 (reference [40]).

NEF **Network Element Function (NEF) Block.** This function, performed in a Network Element, communicates with the TMN for monitoring and control of that Network Element.

QAF **Q Adapter Function (QAF).** This function is provided to connect network elements into a TMN network when they themselves are not TMN-capable. As can be seen from Figure 5.7, an 'M' interface connects to the non-TMN device's Management System (which could be as simple as a command-line interface). The 'M' interface is different for each different piece of non-TMN equipment and does not form part of the TMN standard.

Mediation Function (MF). The Mediation Function, present in a Mediation *MF*
Device was specified to act as a gateway between q_x and q_3 points. Having said this, its precise function is somewhat misty in the standards
documents and its functions in practice overlap those of the Operations
System Function (see below).

Operations Systems Function (OSF). This provides the logical entities rep- *OSF*
resenting the higher management layers described above: the Business
Management, Service Management, Network Management and Network
Element Management Layers. It is with this function that the network
operator, sitting at a workstation and using the Workstation Function, primarily communicates. The device that the Operating Systems Functions
run on is known as the Operations System and, in most networks, is a
reliable set of powerful computers.

Workstation Function (WSF). This normally runs on some form of worksta- *WSF*
tion and is the function which interfaces between the TMN and the network operator.

B-ISDN interfaces with the TMN using the Q interface protocols, as defined
in M.3020 (reference [41]), implemented at the q reference points (q_x and q_3).
The Workstation Functions interact with the TMN through the F interface at
the f reference points. Note again that, should the Workstation Function and
the Operations Systems Functions be implemented in the same physical device,
then no F interface is needed and the f reference point is undefined.

5.6 Interim Local Management Interface   *ILMI*

5.6.1 Structure

As the name suggests, this interface was designed as a temporary measure
to enable some level of management control over ATM UNI interfaces. The
temporary measure was considered necessary by the ATM Forum because the
more grandiose standards described above still have many areas designated as
'for further study'. The ILMI is therefore proposed 'for the interim period until
such standards are available'. It allows operators to obtain status information
about ATM devices and links, operators to configure ATM devices and ATM
devices to register their addresses (see page 68).

The ILMI specification (reference [42]) is based on the Simple Network Management Protocol (SNMP) (see RFC1157 – reference [43]) which was origi- *SNMP*
nally designed for the control and monitoring of fairly simple IP-based devices.
SNMP management has two elements: a database holding configuration and
other information about the device and a standard means for an operator to
access it. The database has a hierarchical tree structure and is known as a
Management Information Base (MIB). The MIB sits between the hardware (and *MIB*
software) of the Network Element and the operator. Changes in the Network

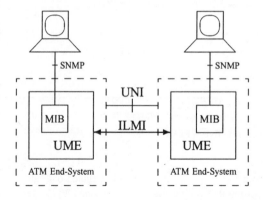

Fig. 5.8 *ILMI Application*

Operation name	Operation
GET	Retrieve specific MIB information
GET-NEXT	Walk the MIB and retrieve the next item
SET	Modify MIB information
TRAP	Report extraordinary event

Table 5.5 *SNMP operation types*

Element's status (links failing or recovering, high error rates, etc.) are reflected in some device-specific manner into the MIB. The operator can then interrogate the MIB using standard commands (see Table 5.5). The operator can also register with the MIB to receive notifications of significant events, *TRAPs*, without having repeatedly to query the MIB for these.

UME ILMI builds on SNMP by logically placing a UNI Management Entity (UME) with each UNI (see Figure 5.8). This entity contains a MIB for the UNI. If a device supports multiple UNIs then there will be a UME for each UNI and if two devices are connected by an UNI, as shown in Figure 5.8, then there are two UMEs associated with the interface: one at each end of the link. ILMI communication always takes place between such adjacent UMEs, over a configured VCC: VPI = 0, VCI = 16 by default. The SNMP commands which travel across this interface use AAL-5 encapsulation (and not UDP and IP as is normally used for SNMP).

In addition to receiving commands from adjacent UMEs, an SNMP command can also arrive from a human operator at a network management station as illustrated in Figure 5.8.

For more details of the type of information held within the MIB, see Table 5.6. The UME is required to handle the SNMP commands listed in Table 5.5

Layer	Item	Note
Physical	Interface index Interface address Transmission type Media type Transmission type Operation status	
ATM	Max. number of active VPCs/VCCs Number of configured VPCs/VCCs Maximum number of active VPI/VCI bits UNI port type UNI version type	
ATM statistics	ATM cells received ATM cells dropped on the receive side ATM cells transmitted	Aggregated over all VPCs and VCCs per UNI
Virtual Path Connection	VPI value Transmit traffic descriptor Receive traffic descriptor Operation status Transmit QoS class Receive QoS class	Config. information such as QoS parameters per end point.
Virtual Channel Connection	Transmit traffic descriptor Receive traffic descriptor Operation Status Transmit QoS class Receive QoS class	Config information such as QoS parameters per end point.
Other	Address registration information	

Table 5.6 *ILMI MIB information*

within specified time limits: reporting events (SNMP *TRAPs*) within 2 s of the event being detected by the UME and responding within 1 s for 95% of requests (*SET*, *GET*, etc.) on single items in the MIB.

The basic idea of ILMI is that information relating to Network Elements be held within those Network Elements rather than within a centralised database at some Network Management site. This technique has become possible over the last few years as Network Elements have become more intelligent and able to handle protocols such as SNMP. The use of the network as its own database (or, equivalently, having a very distributed management system) eliminates the need to keep a centralised database up-to-date as events happen in the network, but the time taken to satisfy operator enquiries is greater as an enquiry has to be sent to the Network Element. This can be obviated to some extent by the Workstation Function caching information but, unless it is refreshed, this can become out-of-date and the ILMI standard calls for information held within the OAM system, and which is older than 30 s, to be flushed.

6
ATM AND OTHER PROTOCOLS

6.1 Introduction

In order to be successful, ATM must be able to interact with protocols already used in networks. In particular, with the growth of the Internet and the increasing importance of Intranets and Extranets, being able to inter-work with the Internet Protocol (IP) is essential.

When ATM has to work with another protocol, the interaction can take two forms:

- It may be necessary to carry information already encoded according to one protocol across an ATM network. In this case ATM treats the incoming information as a stream of bits, has no knowledge of the other protocol and simply breaks the data into cells, transmits these across the ATM network and reassembles them before passing them back to the higher protocol. The disadvantage of this technique, known as encapsulation, is *Encapsulation* that it can use a lot of unnecessary bandwidth: ATM headers are simply added to the headers already on the message, increasing the overhead being carried. An example later in this chapter (see page 97) illustrates 159 octets being transmitted across a network in order to carry 64 octets of useful data. This is not unusual and it is salutary to realise that even 1 octet of user information would, in the example given, be transferred across the network buried in 159 octets. When using encapsulation, an ATM network can be considered simply as a *layer* 2 for the protocol being carried. For this reason ATM is sometimes referred to as a layer 2 protocol.
- It may be necessary to decode the incoming messages from their original encoding and recode them in ATM format. This function is more complex but is used, for example, when a terminal 'talking' Frame Relay is connected to a terminal 'talking' ATM. In general, this technique is slower than simple encapsulation but uses less bandwidth.

In the remainder of this chapter we will consider ATM's interaction with two common protocols: the Internet Protocol (IP) and Frame Relay. Note that, although in this chapter we have discussed pairs of protocols (ATM with IP and ATM with Frame Relay), it is common in practice to find several protocols on top of each other. It is not unreasonable, for example, for IP traffic to be carried across a Frame Relay network (IP over Frame Relay) and for the Frame Relay to be carried across an ATM network (IP over Frame Relay over ATM). IP over IP over Frame Relay over ATM is also not unknown. Of course, at every junction between protocols, additional headers are added: eventually the network is carrying mainly headers, a very uneconomic process.

6.2 ATM and IP

6.2.1 Background

Both ATM and IP are, at the time of writing, moving targets. In particular IPv6 (IP version 6) is becoming standardised. However, almost every installation of IP in the field conforms to IPv4 and, unless we say otherwise, we refer to IPv4 in what follows. Even more volatile than the ATM and IP specifications are the specifications for their interworking, but this chapter will try to capture the state of play at the time of writing.

ATM and IP are strange bedfellows. Almost everything in their philosophies and implementations are in direct contradiction:

- IP carries variable-length packets, with lengths varying from 12 to 65 536 octets; ATM carries fixed-length cells of 53 octets.
- IP 'solves' the matter of Qualities of Service by using very high bandwidth connections; ATM by having sophisticated Quality of Service specifications and using traffic-shaping to enforce them.
- IP is generally *routed*; ATM is switched. This important distinction is explained below (see page 110).
- IP is designed for use on broadcast media and many of the features of IP (for example, ARP, the Address Resolution Protocol) rely on this broadcast nature; ATM is fundamentally a point-to-point protocol, having to simulate a broadcast by using point-to-multipoint or multiple point-to-point connections.
- IP charges are normally based on bandwidth; ATM charges are typically based on usage.
- IP 'connections' (actually TCP connections making use of the IP layer beneath) do not have a defined path through the network; ATM connections have a predefined path.

ARP

Several of these differences reduce ultimately to Quality of Service. For example, the reason for the fixed size ATM cell is to allow parts of frames to be interleaved, preventing a long data frame from delaying a short, delay-sensitive voice frame. The IP and ATM approaches to Quality of Service are like a private motorway where the owner charges cars to use the motorway but, in return, guarantees a certain quality of service – lack of traffic jams, etc. The owner can offer this guarantee either by building an enormously wide road which can never get congested, irrespective of the number of cars using it (the IP way) or by carefully controlling the traffic allowed on to the motorway to ensure that it never overloads (the ATM way).

There is an even more important area of incompatibility between ATM and IP, arising from IP's handling of the situation when a congested ATM network discards a single cell and thereby corrupts an entire packet. This problem is described in more detail below (see page 94), after the main characteristics of IP have been explained.

IP and ATM meet in two important areas: at the desktop and in the backbone (wide area) network. At the desktop, the ATM25 standard has been proposed to bring ATM into the workstation, thus replacing IP. This is a straight ATM application and is not considered further in this chapter. In the wide area network, ATM is expected to carry IP traffic.

Given the fundamental incompatibility between ATM and IP, it is reasonable to ask what is encouraging their marriage. The answer, as indicated elsewhere in this book, is partly fear. To the telephone operators, merging voice and data was supposed to mean bringing data into the voice world, using B-ISDN. To the traditional data equipment suppliers, the concept meant bringing voice into the data world, implementing voice over IP. This distinction is discussed, albeit somewhat melodramatically, in reference [44], where the author contrasts the slow-moving, quality-conscious, voice-oriented carriers (BELLheads – Bell being almost synonymous with telephony in the USA) with the fast-moving data equipment providers (NETheads). One of the weapons used by NETheads (who, incidentally, have issued badges carrying the number 53 crossed out) is the term Cell Tax to describe the 10% header carried on each ATM cell. The negative connotations associated with the term *Tax* have been well exploited by this group which seems to have forgotten the headers carried within IP frames, which can often exceed 10% of the payload.

Another driver for the IP/ATM marriage is the rise of corporate Intranets and Extranets. An Intranet is an internet which is private to a company and an *Intranet* Extranet is an extension of an Intranet to include access to the networks of *Extranet* the companies with whom the Intranet owner does business. By linking together the networks at the various company sites and making use of Internet technology, a company can cut much of the paperwork associated with distributing documents, publicising events, handling employees' expense vouchers, etc. Increasingly, large companies are moving to Intranets and are looking to the carriers who transport the traffic between their sites to manage their Intranet for them. Carriers are therefore being required to carry IP traffic for long distances over their Frame Relay and ATM backbone wide-area networks. In the past this would have been achieved by connecting IP routers, one physically attached to each LAN, with leased lines as shown in Figure 6.1.

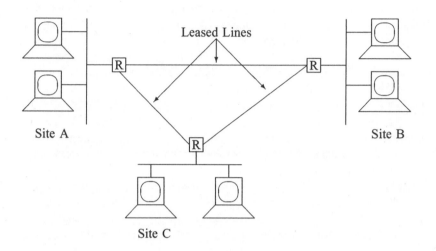

Fig. 6.1 *Historical LAN Interconnection*

This solution is costly to the end-user because leased lines are charged on a *bandwidth × time* product, not on usage. It is also inconvenient because the units of bandwidth that can be purchased are typically widely spaced (64 kb/s, 2 Mb/s, 34 Mb/s in Europe or 56 kb/s, 1.5 Mb/s and 45 Mb/s in the USA: see Table 8.1) and a company cannot select any intermediate speed.

It is also a difficult solution for the carrier to implement when a large number of LANs is involved: the choices are either to proliferate the leased lines (an n^2 problem) or to introduce a hierarchical system where packets pass through a number of intermediate routers. In addition to being costly to the end-user, leased lines are also less than perfect for the carrier – with the liberalisation of telecommunications markets around the world, the sale of leased lines has become a commodity market where the only real differentiator is price. This is a part of the market in which it is difficult for many carriers to compete. Offering an alternative transport mechanism with guaranteed Qualities of Service, together with the management of the end-customer's network is a very attractive idea – this is known as *moving up the value chain*. Another disadvantage of leased lines for the carrier is that the bandwidth which the end-customer is purchasing must be available for that customer at all times; the carrier cannot reuse the bandwidth when the end-user is not using it.

The carrier's solution to the problems associated with leased lines is to offer to carry the LAN-to-LAN traffic across a wide-area ATM network as shown in Figure 6.2. Initially Permanent Virtual Circuit (PVC) connections with CBR Quality of Service could be offered to replace the leased lines directly. This would, at least, allow the carrier to offer any particular bandwidth that the customer required and use any under-utilised bandwidth for other customers.

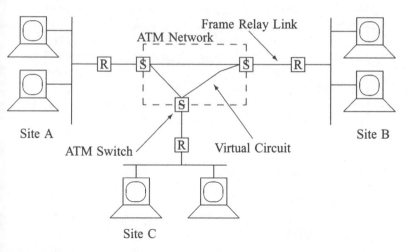

Fig. 6.2 *Contemporary LAN Interconnection*

Subsequently the links could become Switched Virtual Circuits (SVCs), established only when required, and offering ABR, VBR (real-time or non-real-time as appropriate) or UBR Qualities of Service. With all of these options available, a carrier would no longer be selling the same commodity as its competitors – the carrier could now differentiate its service other than on price alone.

Given, therefore, that IP must be carried across an ATM network, various standards have been written to allow this to be done. These can be classified as follows:

- Application-Driven: AREQUIPA
- Network-Driven
    - Classical: Classical IP, Multiprotocol Encapsulation Over AAL-5, LAN Emulation (LANE), Multiprotocol Over ATM (MPOA)
    - Label-Switching: IP Switching, Switching IP Through ATM (SITA), Aggregate Route-Based IP Switch (ARIS), Cell-Switched Routing (CSR), Tag Switching, Multiprotocol Label Switching (MPLS)

Of these, AREQUIPA is the odd one out. It is a proposal from the Swiss *AREQUIPA* Federal Institute of Technology contained in RFC2170 (reference [45]) for a protocol which would allow applications using IP to create an ATM Virtual Circuit to carry their traffic. All of the other techniques listed above are transparent to the application, being offered by the network without the application's knowledge.

All of these techniques try to solve the problem of efficient IP transport from the point of view of the ATM network; there are also significant developments underway to enhance IP itself to give it many of the beneficial characteristics of an ATM network (e.g. different Qualities of Service negotiated at call establishment). These purely IP-oriented solutions are not considered further in this book but the interested reader is referred to descriptions of the Resource ReSerVation Protocol (RSVP) and the Internet Stream Protocol (STII) in references [46] and [47] respectively.

RSVP
STII

The remainder of this section consists of a brief description of IP, to put the discussion into context, followed by a description of the network-driven techniques listed above.

6.2.2   The IP Protocol

There are many good books available describing IP in detail (for example, reference [48]) and no attempt is made here to give more than the outline required to understand the implications of IP for ATM.

IP was invented to be the network layer (layer 3) protocol running on Ethernet and Token Ring LANs. In these broadcast environments it is easy for one IP device to announce its presence and to discover the identities of its neighbours. When the LAN is replaced by an ATM network, broadcasting is lost and has to be simulated in some way.

IP is an unreliable protocol, discarding messages when necessary without attempting retransmission and without warning the transmitting program. Most applications which require a reliable protocol across a LAN therefore make use of the Transmission Control Protocol (TCP) transport layer (layer 4) protocol which sits above IP and which automatically retransmits packets lost by IP. Some applications, which themselves handle retransmission, make use of the unreliable User Datagram Protocol (UDP) transport layer protocol. Not needing to be reliable, UDP is significantly simpler and faster than TCP and, unlike TCP, is not connection-oriented − TCP builds a connection before transmitting information, UDP simply launches each packet into the system without concern about the readiness, or even the existence, of the recipient.

TCP

UDP

IP originated in the LAN environment where each device is identified by a unique 48-bit address known as its MAC address. This address is globally unique and is built into each device at the time of its manufacture. This is a classic example of a flat (nonhierarchical) addressing scheme and, as such, cannot be used directly by applications as they would have to be changed every time a device on a LAN was replaced. Instead, the MAC addresses are currently mapped to 32-bit IPv4 Addresses (which are themselves often mapped to meaningful names). These 32-bit addresses are the ones with which everyone using a LAN is familiar: normally written as four decimal numbers in the range 0 to 255 separated by dots (for example 47.1.192.89). There is a sophisticated structure in these addresses which allows a hierarchical scheme to be implemented.

MAC

IPv4

IP addresses which appear outside of purely private networks need to be globally unique and this is achieved by having one organisation, the so-called

InterNIC – the Internet Network Information Centre – allocate all of the ad- *InterNIC* dresses worldwide. It is the necessity for global uniqueness which has caused IP addresses to become a bottleneck; with the unpredicted recent growth of the Internet, the addresses are becoming exhausted.

A new IP standard has been proposed to solve this exhaustion and the problem of IP not supporting different Qualities of Service. This standard, known as IPng *IPng* (IP, next generation) or IPv6 (IP version 6), extends the IP address from 32 bits *IPv6* to 128 bits. Huitema, in reference [49], estimates that this will provide at least 1 564 addresses for each square metre of the earth's surface (and possibly as many as 3 911 873 538 269 506 102 per square metre). It is very dangerous in the computing or telecommunications industry to predict that any number is future-proof but it is probably fair to say that the 128-bit IPv6 address is likely to be adequate for quite a while, at least until we colonise Mars.

In practice, not all companies register their addresses with the InterNIC and, assuming that their networks will always be private (i.e. without connection to the global Internet), use arbitrary IP addresses. Carriers now offering to manage these Intranets for the owners are finding the overlapping of IP addresses between different companies causing serious problems, particularly in the area of network management.

One crucial concept regarding the interconnection of LANs is how different LAN segments (Ethernets, Token Rings, etc.) may be connected together. Four different techniques, each corresponding to a different level of the protocol stack, are used:

- Two LAN segments may be joined together at the Physical Layer (layer 1) by using a repeater. This device joins the two segments together elec- *Repeater* trically without attempting, in any way, to interpret the signals that it is transmitting. Since the repeater is not intelligent there can be no filtering of messages at this point and the two joined LAN segments behave as one segment. This technique is limited normally by the maximum delay allowed on the network.

- Two LAN segments may be joined at the Data Link Layer (layer 2) by a Bridge. This device forwards Link Layer messages without interpreting *Bridge* their contents and is thereby able to operate on any internetwork protocols being used on the LAN segments.

- Two LANs may be joined at the Network Layer (layer 3) by a Router. *Router* A router is a much more complex device than a Bridge since it needs to understand the internetwork layer protocol (IP, IPX, etc.). A router will listen to the packets interchanged on one LAN segment and, where applicable, forward them to another LAN segment. The router can be thought of as being connected to both LANs, probably having a different IP address on each.

- Two LANs may be joined at the application layer by a Gateway. A Gateway is an application program which receives messages from one *Gateway*

LAN and, where applicable, retransmits them onto another.

In what follows, the distinction between bridging and routing becomes very important. Bridging, it should be remembered, does not understand network layer protocols such as IP. In practice, however, these clear theoretical distinctions are being eroded by equipment which offers functionality spanning more than one layer. An additional complexity is introduced by the technique of switching IP traffic.

Traditionally, IP has been a *routed* local area protocol and Frame Relay and ATM have been *switched* wide area protocols. The difference between these techniques revolves around what is carried in a packet's or cell's header. A routed packet will typically carry the address of its destination and a routing decision is made to determine which outgoing link is appropriate for that destination. A switched packet or cell will typically carry some indication of the path that it is following. Switching, then, consists of transferring the packet or cell to the outgoing link for that path. Switching is normally much faster than routing as, in routing, each packet is considered separately whereas in switching the mapping from the input port to the output port is predefined.

Switching is therefore achieved by labelling the packet with a value indicating its ongoing route. This label may be an ATM VC identifier or any other artificial value with meaning only to the local switch and the switch need only examine the label to identify the outgoing route of the packet without consulting routing tables, etc.

TTL

One general disadvantage in switching (rather than routing) IP traffic is that the Time-To-Live (TTL) field in the IP header is not decremented at each switch; normally this field is decremented at each router and is used to detect, and discard, packets which are looping endlessly. One of the advantages claimed for IBM's ARIS protocol (see page 115) is that, although it switches IP traffic, it can handle the decrementing of the TTL. This IP routed/ATM switched distinction is beginning to disappear as companies are coming to market with various ways of switching IP traffic (see the discussions on IP Switching, SITA, ARIS, Cell-Switched Routing, Tag Switching and MPLS below).

One final characteristic of IP important to its transfer across a wide area network is the manner in which a lost packet is handled. At the IP level the handling is simple: IP is an unreliable protocol and it does not get excited about a lost packet, relying on a higher-level protocol to retransmit a copy of

TCP

it. Most connections make use of the TCP protocol to carry out the necessary retransmission and TCP's retransmission characteristics can cause problems for the underlying network.

ftp

When data are being transferred in bulk (for example, when using ftp, the File Transfer Protocol) the source expects an acknowledgement from the destination that transmitted packets have been received. Once a packet has been acknowledged, the source knows that it can discard the copy it has been keeping as no retransmission will be necessary. If a packet is lost *en route* then no acknowledgement arrives and, after a timeout[1], the TCP protocol causes the source to resend (and increases the timeout for next time). If no acknowledgement is again received, TCP waits for the increased timeout before retransmitting and increases the timeout yet again. When an acknowledgement is eventually received, a further timeout occurs before the next packet is transmitted and the timeout is reduced only when a non-retransmitted packet is acknowledged. On a local area network this behaviour of backing off and starting again slowly is very public spirited as it avoids clogging the network when the network is overloaded; on a wide area network it can lead to excessive delays.

It is this behaviour that was alluded to above: ATM networks can aggravate the problem of retransmission and thereby dramatically reduce the TCP throughput. Each TCP packet is broken into a number of ATM cells and an ATM switch is likely to see cells from different TCP packets arriving sequentially. When an ATM switch becomes congested it discards incoming cells until the congestion abates. If this discarding is unintelligent, it is probable that one cell from each of many packets will be discarded. These packets will then not be able to be recreated at the destination ATM switch. This has two dramatic impacts:

- The ATM network wastes significant bandwidth carrying a large number of cells across the network only for them to be discarded at the destination once it is clear that the entire packet cannot be reconstituted.

- The sources of all of the corrupted packets will timeout and retransmit, slowing the TCP throughput of many connections.

This problem, which has been studied in numerous papers (see, for example, reference [50]) can be addressed by more intelligent cell discarding by the ATM switch. If, instead of discarding one cell from each of many packets, the switch discards all of the cells from one packet, both problems are solved, as ATM congestion will be reduced and TCP will only need to retransmit one packet. This technique is known as Early Packet Discard (EPD) and is applicable to *EPD* AAL5 where the underlying frame structure is understood by the ATM switch. The problem of determining how to choose which packet to discard in a fair manner (for example, so that one particular user does not always get his packets discarded) remains an unsolved problem. When EPD is added to an Unspecified Bit Rate (UBR) service the combination is sometimes known as UBR+ (a term *UBR+* possibly coined by *Data Communications* magazine in its April 1996 issue). The performance of UBR+ when carrying TCP traffic compares favourably with ABR in many situations and is easier to install since it needs no additional intelligence in the generating device.

---

[1]This is a gross simplification of the highly complex restart and retransmission of TCP described in numerous RFCs and reference [48], but the principle is correct.

6.3 IP transfer techniques

This and the following section describe standardised and proposed techniques which address the use of ATM in conjunction with internetwork protocols such as IP. The remainder of this section describes the 'traditional' methods of handling IP over ATM:

- RFC1483 is a standard for encapsulating IP traffic within ATM.

Classical IP
- RFC1577 is a standard for so-called 'Classical IP' over ATM. This standard makes use of RFC1483 encapsulation to provide an IP service across an ATM network.
- LAN Emulation (LANE) is a technique which allows a physical LAN segment (an Ethernet or Token Ring) to be replaced completely by an ATM network.
- Multiprotocol Over ATM (MPOA) builds on LANE and Classical IP and is a technique whereby ATM can play the rôle of a router in a network of several LAN segments.

In addition to these standards, various manufacturers have developed their own ways of handling IP efficiently across an ATM network. These are discussed in subsection 6.3.2 starting on page 106.

Before diving into the details of these techniques, it is worthwhile to state explicitly the problem that they are trying to solve:

How can IP frames be transferred across an ATM network without the ATM cells being reassembled into frames at each switch to allow IP routing to be carried out?

The delay inherent in reassembling the IP packets from the ATM cells in order to perform IP routing (at layer 3) needs to be avoided to allow IP/ATM networks to compete with pure ATM and pure IP networks in speed.

6.3.1 Classical techniques

Multiprotocol Encapsulation Over AAL-5

RFC1483

RFC1483 (reference [51]) describes a technique for encapsulating IP traffic for transfer across ATM Virtual Circuits using ATM Adaptation Layer 5 (AAL-5). This technique is similar to using leased lines for interconnection: one or more ATM VCs are established between the routers or bridges on each of the LANs and LAN traffic is simply encapsulated, broken into ATM cells, transmitted and then reassembled into the original packet. While this still simply offers point-to-point connections for the end-customer, it allows the carrier to make better use of the backbone network – reusing bandwidth temporarily not required for other traffic.

Figure 3.13 shows the AAL-5 PDU structure and it is precisely this structure which is used to encapsulate the IP packet.

ROUTED

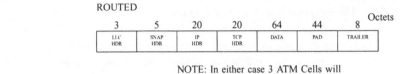

3	5	20	20	64	44	8	Octets
LLC HDR	SNAP HDR	IP HDR	TCP HDR	DATA	PAD	TRAILER	

NOTE: In either case 3 ATM Cells will
be used. Total size is then 159 octets.

BRIDGED ETHERNET

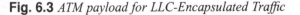

3	5	2	12	2	20	20	64	8	8	Octets
LLC HDR	SNAP HDR	PAD	MAC HDRS	LEN	IP HDR	TCP HDR	DATA	PAD	FCS	

Fig. 6.3 *ATM payload for LLC-Encapsulated Traffic*

RFC1483 defines two modes of working: one where the multiplexing (i.e. the transfer of different streams though one pipe) of different protocols is performed by creating different ATM VCs (known as VC-based Multiplexing) and one where the multiplexing is carried out on a single ATM VC (LLC Encapsulation).

The technique of LLC Encapsulation was included in RFC1483 to accommodate ATM networks on which the establishment of a VC takes a long time or many resources – by establishing a single VC and using it for all of the traffic, continual establishment and breakdown of VCs is avoided.

As in all things in telecommunications, the flexibility of LLC Encapsulation comes at a price, in this case the price of extra overhead within the packet to allow it to be demultiplexed at the destination. Two types of LLC Encapsulation are defined: one for routed and one for bridged traffic. Figure 6.3 shows the respective structures that are presented to ATM AAL-5 as the payload of the CPCS PDU for routed and bridged traffic. As an example, this figure and Figure 6.4 assume that the user wishes to transfer 64 octets of information and, in the case of the bridged protocols, that the medium is an Ethernet. Note that the *PAD* field is used to pad out the data so that the *TRAILER* appears at the end of an ATM frame (i.e. the total length of the PDU and the *PAD* is 8 octets short of a multiple of 48).

The payload structure is much simpler for VC-based multiplexing. Because an individual VC is established for each connection, no information needs to be transferred in the data about the protocol being carried (this having been agreed previously between the two ends: either through signalling or through specification by the user). Figure 6.4 shows the respective structures corresponding to routed and bridged traffic.

As shown in the figures, either of the encapsulations described in this section adds overhead to the information being transferred. Irrespective of the encapsulation, with 64 octets of user data, 159 octets are actually transferred across the ATM network. It is salutary to note that this overhead added to a 64 octet packet is almost 1.5 times the actual useful information transferred and therefore

ROUTED

20	20	64	44	8	Octets
IP HDR	TCP HDR	DATA	PAD	TRAILER	

NOTE: In either case 3 ATM Cells will
be used. Total size is then 159 octets.

BRIDGED ETHERNET

2	12	2	20	20	64	8	8	Octets
PAD	MAC HDRS	LEN	IP HDR	TCP HDR	DATA	PAD	FCS	

Fig. 6.4 *ATM payload for VC-based Multiplexed Traffic*

that the total transmitted unit is almost 2.5 times the 64 octets requested by the
user.

Classical IP   Classical IP

RFC1577   RFC1577(reference [52]) is entitled *Classical IP and ARP over ATM* and was an
early attempt to provide basic IP transmission capability over an ATM network.
    The mapping between an IP address, which is used across the network, and
the physical MAC address of the device is normally performed in an IP network
ARP   by using the Address Resolution Protocol (ARP). This is a very simple protocol
which simply broadcasts to all listening stations a message of the type 'Who
has the physical address corresponding to IP address?' This type of request
cannot be handled easily across an ATM network since ATM links are point-to-
point and broadcast is therefore not possible. Consequently, RFC1577 defines a
device, known as an ATMARP Server, residing on each Logical IP Subnetwork
LIS   (LIS) which accepts address resolution requests and returns the ATM address
ATMARP   corresponding to the IP address. This protocol is known as ATMARP. Note
that each Logical Subnetwork needs its own ATMARP server and that Logical
Subnets have to be connected together with routers as shown in Figure 6.5.
    Of course, the ATMARP Server can only reply to address resolution requests
once it has established a table of IP/ATM address pairs. It establishes this table
inATMARP   by using the inATMARP protocol to request the IP address of each ATMARP
Client which contacts it. In order to keep this information fresh, each ATMARP
Client (the process running in the end-stations on the network) is required to
establish an ATM VC to the ATMARP Server at least every 20 minutes and
update its own ATM/IP address pair in the Server.
    Once an IP address has been associated with an ATM address (private or
E.164 – see page 67), an end-station establishes an ATM VC across the ATM

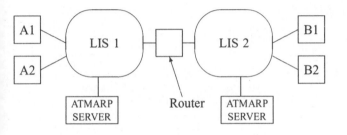

Fig. 6.5 *Logical IP Subnets joined by a Router*

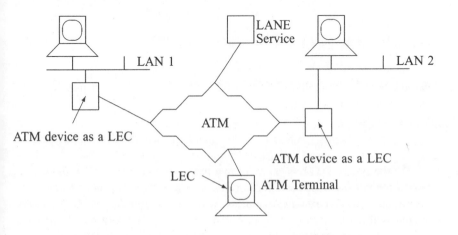

Fig. 6.6 *A LANE network*

network to the destination (unless these VCs have been permanently configured) and then uses RFC1483 encapsulation to transfer the IP packets.

To save unnecessary transmissions, an ATMARP Client itself keeps a cache of IP/ATM address pairs which it receives from the ATMARP Server and at least every 15 minutes it reconfirms each entry by transmitting an inATMARP request to the Client at the other end of each of its ATM VCs.

LANE   *LANE*

LAN Emulation on ATM, or LANE, is the low-level (MAC level) simulation of a LAN segment by an ATM network; that is, LANE bridges at the Link Layer.

The LANE standard (see references [53] and [54] which contain the definitions of the UNI and the NNI respectively) is designed to allow an ATM network to replace a single segment of a physical LAN (either an Ethernet or a Token Ring) and to allow devices equipped with ATM interfaces to communicate with

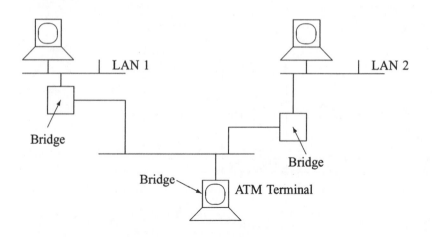

Fig. 6.7 *LAN Network equivalent to Figure 6.6*

devices on a LAN.

Application programs running on the LANE device should find the LANE indistinguishable from a real Ethernet or Token Ring LAN. The interaction between the IP and ATM stacks is shown in Figure 6.8. On each ATM device *ELAN* participating in the emulated LAN (ELAN), a program known as the LANE *LEC* Client (LEC) runs. This program is the interface between the MAC layer and the emulated LAN. In this position, it is the LEC which pretends to the higher levels that they are running over an Ethernet or Token Ring network.

Figure 6.6 shows the manner in which LANE could be used and Figure 6.7 shows the equivalent logical configuration with a real LAN in place of the Emulated LAN. A comparison of the two diagrams will show that, in the latter, the LECs have metamorphosed into Ethernet Bridges and the LANE service has disappeared.

Of course, emulating a LAN segment with an ATM network has certain problems:

- A real (non-emulated) LAN segment, whether Ethernet or Token Ring, is a broadcasting medium and the IP protocols rely on this mechanism, for example, to convert IP addresses to MAC addresses (see the description of ARP above).
- A further level of address translation needs to be carried out somewhere. On a real LAN segment, the IP address is translated into a MAC (hardware) address before transmission. On an ATM network it must be converted to an ATM address.

To handle these problems, three servers are defined which must also be con-
LECS nected to the ATM network: the LAN Emulation Configuration Server (LECS),

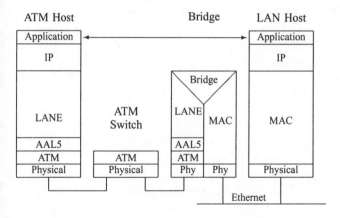

Fig. 6.8 *IP and ATM Stacks with LANE*

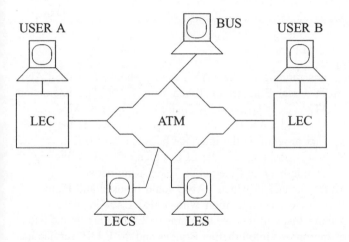

Fig. 6.9 *Example ELAN*

the LAN Emulation Server (LES) and the Broadcast/Unknown Server (BUS) . *LES*
Jointly these three servers are known as the *LANE Service*. *BUS*

In order to illustrate the use of these servers at a high level, it is probably easi- *LANE*
est to work through an example of a new device attaching itself to an ELAN and *Service*
then starting to transmit IP traffic (see Figure 6.9). Once the general approach
is clear, we will take each step in more detail. Note that the three servers in the
figure are shown as workstations and as distinct devices. In practice some manufacturers are implementing the LANE Service as a single workstation, others as
part of the ATM switches themselves. Again, the standard does not require that
the servers be in separate devices and, in many implementations, they appear as
three programs running on the same machine.

- When the ATM device containing the LEC first comes up (say for user A in the figure, we will assume that user B's LEC is already up and running), the LEC must 'register' its presence so that other devices can contact it. It does this by sending a message to the LAN Emulation Server (LES). Unfortunately, at this time the LEC does not know the address of its LES and is therefore unable to contact it. It first builds a Virtual Channel (VC) across the ATM network to its LAN Emulation Configuration Server (LECS). The obvious question, *if it doesn't know the address of its LES, how does it know the address of its LECS?*, will be answered in the detailed description below.

- Once the LEC has built a connection to its LECS, it sends a message asking for the ATM address of its LES. The LECS, after validating the request, returns the LES's address and so the LEC is able to build a VC to it.

- Once a VC exists from the LEC to the LES, the LEC is able to register as part of the Emulated LAN 'organised' by the LES. In registering, it sends the LES details of its ATM address, LAN type (Ethernet or Token Ring) and, optionally, its MAC address.

- At this point the LEC is a member of the Emulated LAN but it cannot broadcast as it would on a real LAN. The Broadcast/Unknown Server (BUS) serves this rôle by establishing a point-to-multipoint (and therefore unidirectional) Virtual Circuit to every LEC on the Emulated LAN. Before the LEC can be a full member of the Emulated LAN it must find the address of its BUS and it does this by asking the LES, to which it has a VC, to resolve the all 1s address (0xFFFFFFFFFFFF). The all 1s address is conventionally the broadcast address on a LAN segment and so the LES returns the address of the BUS.

- The LEC is now ready to be a full device on the Emulated LAN. We will assume that some form of higher-layer packet is now sent from user A to user B (this might, for example, be a TCP packet to establish a connection for telnet traffic). When this packet arrives at A's LEC, that LEC will check its tables to see whether it has an ATM VC to the destination indicated by the MAC address in the packet. In this case, as this is the first packet, it doesn't, so it sends the packet across its pre-established VC to the BUS. The BUS will transmit it onward and it will eventually arrive at user B's LEC.

- This process of sending all packets to the BUS for transmission could continue indefinitely (and this would simulate closely an actual Ethernet segment where all packets are broadcast to all stations) but this is an inefficient use of the ATM network. To avoid this inefficiency, user A's LEC, using its pre-established VC, asks its LES for the ATM address of the device supporting the MAC address that it has just seen. The LES returns the ATM address of user B's LEC and user A's LEC establishes a

VC directly to it. While this is happening more packets may be received for transmission but they are all forwarded to the BUS.

- Once an ATM VC is established between user A's LEC and user B's LEC (known as a Data Direct Connection or Cut-Through), further packets are *Cut-Through* sent directly instead of being re-directed through the BUS. By this means a *Connection* fast track has been created. Note that this means that, during the transition period, some later packets taking the fast route may arrive before earlier packets taking the route via the BUS. It is the responsibility of the higher-level protocols at the destination either to reorder the packets, or to discard those arriving out of order and cause a retransmission.
- When the LECs notice that no traffic has made use of the Cut-Through Connection for some time, they tear down the ATM VC.

In summary, the purpose of LANE is to build direct ATM connections between end points (LECs) when traffic is flowing and to tear these connections down when traffic stops flowing. This allows the high speed of the ATM network to be used without the need to build a complete mesh of ATM connections. Of course, even when Cut-Through Connections have been established, there will be packets that require broadcasting and these are still directed to the BUS.

One point of detail was deferred from the high-level description above: the resolution of the chicken-and-egg problem of a LEC not knowing the ATM address of its LES and retrieving this information from the LECS. How does the LEC know the address of the LECS? The answer is simple: it will probably be preconfigured through the OAM system. If it is not preconfigured then the LEC is allowed to make use of a *well-known* ATM address. If this also fails then the LEC will fall back to using VPI = 0, VCI = 17.

It can be seen that the operation of an Emulated LAN depends crucially on the availability of the various servers (collectively known as the LANE *LANE* Service). The LANE specification therefore deals with the operation of the *Service* system following a server failure.

The primary advantage of LANE is also its disadvantage: it can be implemented in a network very simply because no changes are required to any software already using the LAN other than the low-level drivers; to the higher-level software an Emulated LAN is just like any other LAN. This is a disadvantage because the application programs are unable to make use of any of the ATM Qualities of Service.

Multiprotocol Over ATM (MPOA)

LAN Emulation (LANE) is a suitable protocol for replacing a single LAN subnet with an ATM network but it does not allow intersubnet communication unless routers are colocated, physically or logically, with each LEC. The Multiprotocol Over ATM specification (see reference [55]), released in mid-1997, extends *MPOA* LANE to support traffic between subnets by including a Network Layer routing function. MPOA relies on LANE to provide intrasubnetwork packet delivery ar on the Next-Hop Routing Protocol (NHRP) to provide the information needed t establish ATM VCs between subnetworks. At the time the MPOA standard wa released, the NHRP standard had not been published and the entire draft NHR standard (reference [56]) therefore had to be copied into the MPOA standard t make it complete.

The basic principles of MPOA are:

- To pass any traffic destined for the local subnetwork to the underlyin, LANE system for direct delivery.
- To pass traffic destined for another subnetwork to a centralised server while detecting 'flows' of traffic and, when a flow is discovered, to buil a direct ATM connection to the destination (a *Cut-Through* or *Short-Cu* Connection) so that further traffic may bypass the server.

Note that, in some networks, security issues may actually prohibit the estab-lishment of an end-to-end ATM connection.

The idea of a traffic 'flow' is crucial to MPOA. This concept is not only relevant to carrying IP over ATM, even within the IP community the concept of a 'flow' has become important in recent years (in particular in IPv6). Early drafts of the MPOA standard left it to individual implementations to define the concept of a flow so that it could be used as a possible differentiator between equipment manufacturers. The final draft version of the specification, however, gave a very clear definition but the first formal issue hedges its bets, repeating the clear definition as a default but stating that 'Other mechanisms (e.g. RSVP) may be used in specific cases to override this mechanism'.

Parenthetically, it may be noted that equipment manufacturers are ambivalent about standards: a manufacturer with a near monopoly in an area will try to resist the introduction of standards which would allow other manufacturers to interface to its equipment, but when trying to break into a market, manufactur-ers are naturally very enthusiastic about standards. The definition of a 'flow', for example, does not significantly affect the interoperability of equipment but would allow a supplier to produce a more efficient product than its competitor. It is one, therefore, where it would be advantageous for established manufac-turers to oppose a standard definition. The standards bodies themselves are, of course, mainly composed of representatives of the manufacturers.

The issued MPOA specification gives a default definition of a 'flow' as being when a predefined number of packets (set to a default of 10) is sent from one particular source to one destination within a given time (set to a default of 1). This is clearly a crucial matter; it would be wasteful to set up an ATM connection only to find that, by the time it was established, the traffic flow had stopped. If the definition of 'flow' is left to individual implementors, it is likely that parameters other than simply packets per second will be used: possibly with attempts to detect the type of traffic (telnet, ftp, rlogin, etc.).

Figure 6.10 (which is taken from reference [55]) shows the underlying strategy

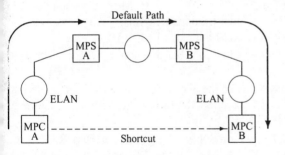

Fig. 6.10 *MPOA traffic paths*

of MPOA. Suppose that MPOA Client A (MPC A) in that figure is an ingress *MPC* to the network and wishes to transfer IP frames to MPOA Client B (an egress point) and suppose further (as indicated in the figure) that MPC A and MPC B belong to different IP subnetworks. As the source and destination are in different subnetworks, LANE cannot be used to make the transmission efficient.

Somewhere, however, on MPC A's Emulated LAN there must be an MPOA Server (MPS) which serves the rôle of a Router. Co-located at the MPS are: *MPS*

- a LANE Client (LEC) to allow the MPS to take part in its Emulated LAN,
- an IP router holding routing tables and tables to map MAC addresses to ATM addresses (in a similar manner to a LANE Server),
- a Next-Hop Server to allow the MPSs to communicate with each other using the Next-Hop Routing Protocol (NHRP).

Although these three functions are logically colocated, in practice they may either be performed by a single processor or be spread across different processors which may themselves be geographically separate.

The sequence of events is then:

- Having received the first packet destined for MPC B, MPC A forwards it (by bridging) to MPS A since it has no information about how it could otherwise reach its destination.
- MPS A is able to route the packet to MPS B which, in turn, forwards it to MPC B. All traffic between MPC A and MPC B could follow this path, but it is clearly inefficient if both parties have access to a common ATM network.
- If the source continues to send frames, either MPC A or MPS A will detect a 'flow' (see above) of information between the two points. If MPS A detects it, then it commands MPC A to start the process to establish a shortcut. If MPC A detects the 'flow' then it starts the process autonomously.

- MPC A sends an *MPOA Resolution Request* to MPS A asking for the information it needs to establish a shortcut Virtual Channel across the ATM network (primarily the ATM address of MPC B).

- MPS A translates the request into an NHRP Resolution Request which it passes to MPS B.

- MPS B now prepares MPC B for the Virtual Channel it will be receiving and then sends an NHRP Resolution Reply back to MPS A, containing the ATM address of MPC B.

- MPS A uses the NHRP Resolution Reply to generate an MPOA Resolution Reply which it passes to MPC A. MPC A can then establish a (switched) Virtual Channel directly to MPC B and packets need no longer be routed through the MPSs.

Once the 'flow' ceases, the Virtual Channel is broken down. In the final draft version of the MPOA specification, a default algorithm was given for detecting the end of a flow (no packets for a very long time (20 minutes was suggested)) but this has been removed from the final version of the specification and is now left to individual manufacturers.

There are many tutorial papers on MPOA appearing in the technical press but reference [57] is particularly good, covering both LANE and MPOA.

6.3.2   Label-Based IP Switching

LANE, MPOA and NHRP are standard protocols defined by the IETF. While the IETF has been working on these standards, various equipment manufacturers have been proposing alternative methods of passing IP (and other Internetwork protocols) efficiently over an ATM network.

Before reading this section, it is perhaps worth restating the problem (see page 96) that all of these techniques are trying to solve:

How can IP frames be transferred across an ATM network without the ATM cells being reassembled into frames at each switch to allow IP routing to be carried out?

The criticism that has been levelled at the LANE and MPOA protocols is that they don't address the basic inconsistencies of the IP and ATM routing protocols: they simply put one on top of the other. This leads to duplication of function in the network and additional complexity in its OAM (fault isolation when a failure occurs, etc.). It is as if ATM and IP are fighting each other.

A selection of the various proposals designed to resolve these issues is described below – some have great technical merit, some appear to have been proposed to give a commercial advantage to a particular manufacturer and some appear simply to be 'spoilers'. The reader might like to apply this informal classification to the techniques described.

The non-MPOA solutions proposed by the different manufacturers can all loosely be called 'Label Switching'. The basic idea is that somewhere around the edge of the network, possibly within the source host itself, a label of some sort is allocated to cells forming part of a flow of information between a particular source and destination. Intermediate switches then do not need to reassemble IP packets, they need only read the label and forward the cell without interpretation or reassembly. The methods differ in the means that they use to determine the label:

- Data-driven methods are like MPOA in that they try to detect flows of information and, when the flow has been detected, associate a label with the flow.
- Topology-driven methods associate labels with particular paths through the network and all traffic following a particular path at the same Quality of Service uses the same label.

In general, data-driven methods of allocating labels suffer from some form of delay when a new flow starts – it takes time for the system to detect that a flow is happening and for the label to be assigned – and from the implicit characteristic that short-lived traffic exchanges (which are not *flows*) have to follow some slower route. Topology-driven methods, on the other hand, suffer to some extent from having to allocate fixed amounts of bandwidth through the network before it is known how much traffic will follow a particular route. Typically, a topology-driven algorithm will require more VCs to be established since the most effective topology-driven network is one with a complete mesh of VCs. It is also argued, primarily by exponents of topology-driven methods, that the data-driven techniques are less robust and stable in the presence of changing traffic patterns.

Data-driven techniques clearly work best for ftp and telnet traffic where there are clearly defined long-lived flows. For short-lived exchanges all of the traffic will be passed through the slower route. As access to World-Wide Web pages becomes a predominant use of networks, there appear to be two contradictory trends: TCP sessions are generally becoming shorter but the increased use of proxy sites is making TCP sessions for commonly accessed pages longer. There is some evidence, published by Ipsilon and backed by Toshiba, that around 80% of packets and 90% of bytes passing across a typical network are not part of such short-lived exchanges but are instead part of a flow (see reference [58]). It is interesting that these two companies are proposing data-driven standards.

In a way, MPOA and LANE can be considered Label Switching standards: the label in this case being the Virtual Channel Identifier.

The IETF, realising that MPOA has not been uniformly accepted, has itself started a standards effort to accommodate all of the various Label-Based options: this is known as MPLS (Multiprotocol Label Switching) and is discussed in its own subsection below.   *MPLS*

This abundance of methods for carrying IP over an ATM network may lead the reader to wonder what the major differences between them are and what criteria can be used to differentiate and judge them. One major discussion at present

	IP Switching	CSR	ARIS	Tag Switching
Champion	Ipsilon	Toshiba	IBM	Cisco
Mapping	Data-driven		Topology-driven	
Inter-switch Protocol	IFMP [59]	FANP [60]	TDP [61]	ARIS [62]
Number of VCs needed	Number of layer 4 flows	Number of layer 3 flows	Number of egress routers	Number of routes

Table 6.1 *Label-Switching techniques*

revolves around how well each behaves in large networks – networks which are geographically large, networks with a large number of users and networks connected to a large number of other networks. Some techniques, for example, require a large number of Virtual Channels to be established, some require a very high rate of setting up Virtual Channels. Another differentiator is how much use each can make of ATM's underlying Qualities of Service: whether these are exploited by associating different types of IP traffic with different Qualities of Service.

The next few sections deal individually with some of the most common existing proposals and the IETF's MPLS. It is clear that commercially only a few, or even one, of these protocols can survive in the long term. It is unclear at the time of writing which this will be. Table 6.1 gives a summary of the four main proposed techniques.

A useful comparison of Toshiba's Cell-Switch Routing and Cisco's Tag Switching is given in reference [63].

Tag Switching

Tag Switching

Tag Switching is a topology-driven technique devised by Cisco Systems Inc., a manufacturer best known for its IP routers. Of all of the techniques described in this chapter, Tag Switching is probably the best known and the most refined. An architectural summary is given in RFC2105 (reference [64]).

TIB

As with the other techniques described in this section, the basic principle of Tag Switching is to build a forwarding table, known as a Tag Information Base (TIB), and to forward incoming ATM cells by consulting this table, using the label (or here, Tag) carried by the incoming cell. The TIB is illustrated in Figure 6.11 where a number of tags are shown in the TIB itself, the entries pointing to the information needed to forward the cell.

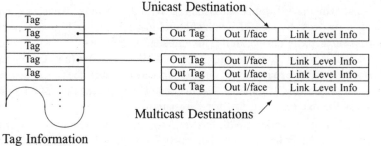

Fig. 6.11 *Tag Information Base*

When a cell arrives at a switch, the switch scans the TIB looking for a Tag equal to the Tag carried in the VCI field of the ATM header (see Figure 2.4). The matching TIB entry points to the Tag to be used in the outgoing cell, to the link to be used and to any necessary layer 2 information for the outgoing link. All the switch needs to do is replace the incoming Tag with the outgoing one and forward the cell to the defined output port. Since Tags are all of equal length, the search for a particular Tag in the TIB is very fast. Note that Figure 6.11 also illustrates how simply Tag Switching can handle multicasting (the sending of the same information to more than one destination) – it is necessary only to have several outgoing Tags (and other information) associated with an incoming Tag.

As the Tag is carried in the VCI field of the ATM header, to avoid clashes with other routing techniques used in the switch (such as PNNI), each routing system must be preallocated its own subset of the 2^{16} possible VCIs.

The meat of Tag Switching is, of course, the means by which the TIB is built in each switch. The protocol proposed for this is known as the Tag Distribution Protocol (TDP).   *TDP*

Each Tag Switch operates at two levels:

- as a router exchanging routing information with its neighbouring routers using a conventional IP routing protocol such as OSPF, the Open Shortest Path First protocol. The results of the routing exchanges are held in *OSPF* the Forwarding Information Base (FIB) as in a conventional router. At *FIB* this level, Tag Switches can interoperate with conventional IP routers, a technique which is perhaps not surprising given Cisco's large installed base of such routers;
- as a switch using the information in the TIB to switch cells.

The purpose of the Tag Distribution Protocol is to build entries in the TIBs of a switch and its neighbours from information in its FIB. Two Tag distribution schemes are proposed, both using TDP:

- Downstream Tag Allocation on Demand. With this technique, each switch sends a message to each of its neighbours identified in its FIB as a next hop towards some destination. This message is effectively a request for the neighbour (the 'downstream' switch) to allocate a Tag (and thereby a TIB entry) for the route to the destination and to return the value of the assigned Tag to the originator. The originator then places this into a TIB entry as the outgoing Tag for that destination.

- Upstream Tag Allocation. With this technique, a switch recognising from its own FIB that it has a point-to-point (i.e. one-hop) link to a particular destination, will spontaneously allocate a Tag, place an entry into its own TIB and then transmit the information 'upstream' to its neighbours who can then update their own TIBs.

Note that both of these techniques are based purely on the topology of the network and neither relies in any way on detecting traffic flows: Tag Switching is a topology-driven process.

In summary, Tag Switching is a technique which overlays the ATM network by using the VCI field in the cell header as a Tag for switching cells without the need to reassemble them into IP frames for routing. Its drawbacks appear to be its close association with one manufacturer (Cisco), which renders it a risk for potential customers, its need to have all ATM switches participating in the Tag Switching process and its failure to take full advantage of ATM's intrinsic capabilities (Qualities of Service, Switched Virtual Channels, etc.). On the other hand it appears to be the most commercially advanced of the label-switching techniques and is likely to be part of the foundations of the IETF's Multiprotocol Label Switching (MPLS) standard.

IP Switching ## IP Switching

IP Switching is a data-driven label technique developed by Ipsilon which requires an IP router to be present in every ATM switch as shown in the upper part of Figure 6.12.

Initially, traffic flowing from user A to user B is passed by each ATM switch across a default Virtual Channel to its colocated router. The router determines the correct outgoing port on the switch and passes the traffic back to the switching matrix.

When a router detects a flow then it assigns a 'label' from its pool allocated to the incoming port and instructs the ATM switch to pass incoming frames with particular header to the router on a predefined circuit. Using the Ipsilon Flow *IFMP* Management Protocol (IFMP) (see RFC1953 – reference [59]), it now sends a message to its upstream node (i.e. the node sending frames to it) containing the label, a flow identifier and a lifetime. This message instructs the upstream node to attach the flow identifier to all packets sent and to transmit them down the ATM Virtual Channel specified in the label. The lifetime field specifies for how long the redirection is valid – if this is not extended within the given time, then communication falls back to the default channel.

This technique by itself speeds up packet processing because incoming packets for a particular flow have already been presorted onto a particular Virtual Channel. The full benefit of the technique, however, comes when the downstream node, working autonomously but presumably detecting the same flows, also applies the technique. Consider Figure 6.12 where three switches are handling a flow from user A to user B. Node Y detected the stream and established a label for the flow from node X. When, however, node Z detected the stream it signalled a label back to Y and Y, realising that it can remove itself from the routing, simply instructs the ATM switch to connect the incoming and outgoing ports.

This technique is being applied by all switches in the chain and eventually all will be working in this fast 'cut-through' mode, thus removing many routers from the link.

The protocol proposed by Ipsilon for use between the router and the colocated switch is known as the Generic Switch Management Protocol (GSMP). This *GSMP* protocol, defined in RFC1987 (reference [65]), was designed for uses wider than IP Switching, being a general-purpose protocol for establishing and releasing point-to-point and point-to-multipoint connections on an ATM switch.

The characteristics of an IP Switching network are that:

- it relies on the detection of 'flows' of traffic and is therefore a data-driven technique
- Virtual Channels are established from switch to switch rather than end-to-end
- it is claimed (by Ipsilon) to be much simpler than MPOA and it certainly requires less configuration by the operator.

IP Switching has a number of limitations, including:

- Special (routing) software is needed in every ATM switch in the connection and the technique therefore cannot be used across an unadapted ATM network.
- The same flow-recognition algorithm must to be implemented in every switch in the connection.
- It is doubtful whether it adapts well to large networks where many thousands of Virtual Channels are required and the rate of Virtual Channels establishment and tear-down may exceed the capacity of many switches. It may also have weaknesses in large networks when a major failure occurs and many ATM Virtual Channels are simultaneously rerouted. This may overwhelm the routers on the new links.
- Having bypassed the routing by building cut-through Virtual Circuits, it can also cause firewall protection to be bypassed. *Firewall*

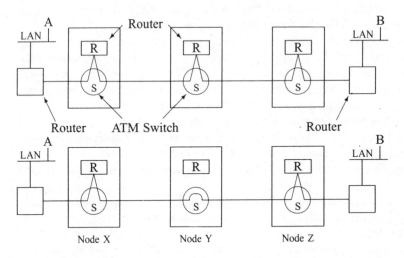

Fig. 6.12 *IP Switching*

- Initially only IP is supported, with other protocols such as IPX having to be tunnelled through IP.

CSR

Cell-Switch Router

Cell-Switch Routing (CSR) is a data-driven technique originating in the Tokyo Institute of Technology and proposed commercially by Toshiba. Being data-driven, it builds its cut-through circuits (known as bypass pipes) based not upon the static topology of the network but upon the traffic flows. At the time of writing, Toshiba and Cisco have proposed a joint Label-Switching scheme which combines the data-driven approach of Cell-Switch Routing and the topology-driven approach of Cisco's Tag Switching – see page 108.

The Cell-Switch Router itself is a device external to the ATM network (rather than being a modification to an ATM switch), connected to the ATM network through a UNI interface. The functions of the CSR are:

- to accept ATM cells on a default VC from another Cell-Switch Router, to rebuild IP packets from these cells and then pass the packets to the router part of the Cell-Switch Router
- to accept ATM cells on so-called dedicated VCs and, without reassembling the IP packets, to forward the cells according to a forwarding table mapping incoming ATM Virtual Path and Virtual Channel Identifiers (VPI/VCIs) to outgoing VPI/VCI.

The first of these, involving reassembly and routing of IP packets, is used for any communications requiring hop-by-hop IP routed paths. VCs set up using

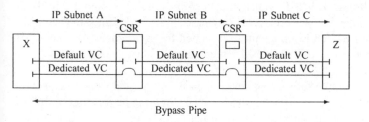

Fig. 6.13 *Cell-Switch Routing*

Classical IP (see page 98) would be candidates for this mode of operation. The second technique, not involving reassembly of the IP packets, is used for flows of traffic.

The underlying idea is to use IP routing to determine the CSRs between which the cells will flow and to use ATM routing between CSRs (within IP subnetworks). Since ATM is routed only within independent (and presumably small) subnetworks, the amount of routing information held by any ATM switch is small. The disadvantage of this technique is that since the ATM routing does not have the ability to see the entire connection end-to-end, the route it follows may be less than optimal – longer, for example, than the one which the NHRP would be able to find. Keeping the two levels of routing separate allows either to change dynamically without affecting the other.

Figure 6.13 (adapted from reference [66]) illustrates the bypass-pipe in a network with two end-user routers (X and Z) which are programmed with the protocols needed to support Cell-Switch Routing (see below) and two actual Cell-Switch Routers. The three ATM networks shown may, in practice, be the same network (in which case the two CSRs are *single arm* routers) or distinct networks.

In Figure 6.13, the originating router (at X) checks each outgoing packet to determine the Virtual Circuit on which the cells constituting the packet should be transmitted. This choice may be determined by the destination IP address, the destination IP address and port number (which determines the type of transfer), the source/destination address pair or, in IPv6, the Flow Label.

If the cells are transmitted on a Dedicated VC, then the intermediate CSRs, will simply forward the cells, without packet reassembly, from the incoming VC to the corresponding outgoing VC. If no mapping is found at a CSR, then the CSR reassembles the packet and passes it to the colocated routing function.

When the cells arrive at the last *Bypass-Capable node*, the packet is reassembled.

Before any of this can happen, the Dedicated VCs (see Figure 6.13) and thereby the Bypass Pipe, must be established. The precise triggers which cause these to be created are not clearly specified in the RFCs, but the protocol which

Toshiba has proposed for the establishment is known as the Flow Attribute Notification Protocol (FANP) which is defined in RFC2129 (reference [60]). In fact, FANP, which is based on Q.2931 (reference [67]), is designed to encompass more than Cell-Switch Routing, and is designed for use on networks other than ATM, but it is as a Cell-Switched Router protocol over ATM that we will consider it here.

In order to build the forwarding tables in each Cell-Switch Router that enable the cells arriving from a particular ATM Virtual Circuit to be forwarded, without IP involvement, to an outgoing ATM Virtual Circuit, some communication is needed between neighbouring (adjacent) Cell-Switch Routers.

Consider the network shown in Figure 6.13 and assume that IP traffic is beginning to flow from X to Z, both of which are FANP-aware devices. As this traffic arrives at X from its hosts, X looks inside the IP packet to decide whether this is a so-called *trigger packet* or not. A trigger packet is one which should trigger the establishment of a Bypass Pipe. As with MPOA, a long-term flow of traffic should cause a Bypass Pipe to be established whereas, for a short exchange, the overhead of establishing the Bypass Pipe is likely to outweigh its worth. MPOA proposes that packets between specific source/destinations pairs be counted to determine whether an exchange constitutes a flow, FANP proposes that X look inside the TCP or UDP header to examine the port to which the packet is bound. A locally configurable table then specifies the ports with which flows are normally associated (such as http, ftp and nntp). Note that this breaks the strict layering rules of the ISO OSI model: we have a layer 3 entity (a router) looking at the layer 4 information in the packet.

trigger packet

Having detected, by fair means or foul, a trigger packet, X then uses a Dedicated VC to the leftmost CSR to send FANP messages. This Dedicated VC may be a Permanent VC (PVC) established as a result of an operator command at start-up or may be a Switched VC (SVC) established as required. In order to reduce the set-up time associated with creating an SVC, X may establish a number of such VCs before they are actually needed.

Having selected a Dedicated VC, X sends a PROPOSE message (part of FANP), proposing a common name for the Dedicated VC (remember that the actual ATM Virtual Circuit Identifiers will be different at each end of the link). This unique identifier is known as a Virtual Connection IDentifier (VCID). The PROPOSE message also contains the IP address of the destination and the receiving CSR looks this up in its routing table to ensure that it is, in its opinion, a valid address.

VCID

Once the CSR (the leftmost in Figure 6.13) has agreed to the PROPOSE message, X sends an OFFER message through the Default VC containing the agreed-upon VCID, a flow identifier (not to be confused with the flow identifier of IPv6) consisting of the source and destination IP addresses and some house-keeping information about the Dedicated VC. Once this has been acknowledged by the CSR, X starts to send IP packets, broken into cells, across the Dedicated VC.

While this FANP communication has been taking place between X and the leftmost CSR in Figure 6.13, the same trigger packet has caused a similar transaction between the two CSRs and between the rightmost CSR and Z. By this means a Bypass Pipe has been created between X and Z. Notice that FANP is a protocol purely between neighbouring nodes but that it is used to create an end-to-end pipe.

Assume now that the flow of IP packets between X to Z ceases. After a timeout period, X detects that the flow has stopped and transmits a REMOVE message to its neighbouring CSR. This may cause the Dedicated VC to be dropped (in the case of an SVC) or the Dedicated VC to be marked as unused (in the case of a PVC).

Thus, the characteristics of Cell-Switch Routing are:

- it is based on Cell-Switched Routers which are not part of the ATM network – they are connected to the network through UNIs
- it separates the IP and ATM routing
- it builds cut-through VCs (known as Bypass Pipes) using a neighbour–neighbour protocol known as FANP
- it determines when to build a cut-through VC based on layer 4 information about the flow.

It should be noted that Cell-Switch Routing, in common with all of the proprietary techniques described here, is not static. At the time of writing, Toshiba is thinking, for example, of adding topology-driven characteristics to the protocol.

Aggregate Route-Based IP Switching

Aggregate Route-Based IP Switching (ARIS) is a topology-driven protocol proposed by the International Business Machines Corporation (IBM). As in other techniques, the idea is to colocate routers with ATM switches to form, in ARIS's case, so-called Integrated Switch Routers (ISRs) and to switch IP traffic rather than route it. *ARIS* *ISR*

Superficially this technique is very simple and revolves around the unique position of Egress Nodes in the network, i.e. switches at the point where an IP flow leaves the network. If each switch in the network could have a preset path to every Egress Node then IP frames could be switched by intermediate nodes towards their point of eventual egress. In a network of N nodes, since there can be at most N Egress Nodes (supposing *every* node to be an egress node), the maximum number of such paths needing to be stored in any node is at most N, and is certainly of $O(N)$ rather than being of $O(N^2)$ as in some other proposed protocols outlined in this chapter. Note that this technique is switching traffic to a particular Egress *Node* rather than a particular IP address. This is particularly effective when there are many IP addresses associated with one Egress Node: all of the flows may be switched using one identifier at each ISR and one entry in the forwarding table.

Figure 6.14 shows a stylised view of the paths through a number of Integrated Switch Routers to a single Egress Node. This typically forms a tree as shown in this diagram and once the traffic for any particular Egress Node has reached an intermediate node such as node A, it can be combined with traffic from other nodes also destined for the same point.

The ARIS protocol defines the way in which each node learns of routes towards the Egress Nodes. Once an ISR has identified itself as an Egress Node, it sends messages to its 'upstream' ISRs (Nodes A and B in Figure 6.14) which contains an 'Egress Identifier', for example the IP address prefix of the destination, informing them of the paths for which it is an Egress Node. When an ISR receives a message of this type it checks to ensure that no routing loops would be introduced by this new route and then adds it to its forwarding table before passing it on to its own upstream nodes, having replaced the Egress Identifier with its own identifier. By this means, information about paths to each Egress Node is disseminated upstream through the network to each ISR. Each ISR can then build a forwarding table accessed by the identifier and giving the path to the next node. This, of course, is happening from each Egress Node until each ISR is a member of a tree rooted at each Egress Node. Once the trees have been established, the ARIS protocol calls for regular 'keep-alive' messages to be sent upstream to cause the forwarding table entries to be refreshed; if the keep-alive messages fail to arrive then it is assumed that the route has failed and the entry is flushed from the ISR's forwarding table.

Instead of this process of dissemination from the Egress Nodes, it is also possible to predefine paths using the ARIS protocol. This could be used, for example, in cases where the source needed to ensure that its messages passed over a particular path.

Once the forwarding table has been established, when an ISR receives an IP frame it uses the identifier contained within it to access its forwarding table to determine the next hop along which the frame needs to travel towards its ultimate destination, an Egress Node.

VC Merge

ARIS will work in different types of network, in particular ATM and Frame Relay networks. In ATM networks, the concept of Virtual Channel Merging, although not supported on many ATM switches, is particularly appropriate to ARIS since it maps very well onto the merging nature of traffic approaching an Egress Node. If switches do not support VC Merging then Virtual Path Merging can be used instead – a tree of Virtual Paths routed at an Egress Node is created carrying unique Virtual Circuits from each ISR to the Egress Node.

VP Merge

IBM Corporation believes that ARIS has several advantages over the other label-based techniques outlined in this chapter. These include:

TTL

- The Time-To-Live field within an IP frame can be correctly updated. The Ingress Node (the first ISR that a frame meets) knows how many steps will be involved in passing the frame to its Egress Node and can

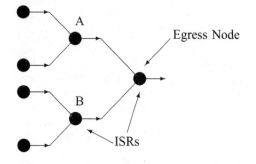

Fig. 6.14 *Aggregate Route-Based Switching Tree*

therefore reduce the Time-to-Live field by the appropriate amount when it is handling it.

- It makes use of any VC Merging capability of the ATM switches and thereby keeps the number of VCs to $O(N)$.
- It is not particular to a type of underlying network, being equally applicable to an ATM or Frame Relay network, and is not particular to the Network Layer Protocol being carried (e.g. IP or IPX).
- The ISRs can coexist in a network with conventional routers and switches – when converting an existing network to ARIS there is no need to convert all routers and switches at the same time.

Switching IP Through ATM

SITA is a proposal from Finnish Telecom intended more as an illustrative ex- *SITA* ample of how the problems of VC merging can be solved than as a practical commercial product. The basic technique is to use the Virtual Path Identifier (VPI) inherent in the ATM protocols to be a 'label' for the route to a particular destination. Thus, Virtual Channels arriving at a switch destined for a common node would be multiplexed onto the same Virtual Path (VP). In this way it can be considered similar to ARIS (see above).

The fundamental problem with the SITA technique for practical applications is that the VPI space in ATM cells at the UNI is only 8 bits (see Figure 2.4 on page 29). This obviously limits the usefulness of the technique to very small networks.

Multiprotocol Label Switching

Multiprotocol Label Switching is the term applied by the IETF to its attempt *MPLS* to bring the general techniques of IP Switching, Cell-Switched Routing, Tag

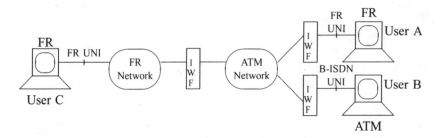

Fig. 6.15 *Interworking Frame Relay and ATM*

Switching, ARIS and SITA together in a standard. The standards work is incomplete at the time of writing but is addressing questions of scalability within the network and support for both unicast and multicast streams.

6.4 ATM and Frame Relay

Frame Relay and ATM are both connection-oriented protocols and both originate in the telecommunications rather than the data industry; their marriage is therefore happier and more comfortable than the ATM/IP marriage described above. The standards required to specify how Frame Relay information should be carried over an ATM network are also much better defined than the equivalent standards for the IP.

Frame Relay is, in many ways, the precursor of ATM, being a lightweight protocol for switching frames (of variable length) rather than cells (of fixed length). ATM has followed so closely on the heels of Frame Relay that, in many markets, carriers had only just begun implementing their Frame Relay networks when ATM arrived.

There are several good references on Frame Relay (for example [68] which is a very short, high-level introduction to Frame Relay and [69] which gives a much deeper introduction to the relevant standards) and no attempt is made here to give an in-depth tutorial on the subject. Suffice it to say that Frame Relay shares many of the characteristics of ATM, including defined User–Network and Network–Node interfaces.

As many Frame Relay networks have been installed and as many Frame Relay devices exist in subscribers' premises, it is essential for ATM's introduction that Frame Relay and ATM work together at least at the following levels as illustrated in Figure 6.15:

- A Frame Relay device should be able to communicate with another Frame Relay device using a B-ISDN network for communication but without either device being aware of the intervening B-ISDN network. This is illustrated in Figure 6.15 by user C communicating from its Frame Relay workstation, over Frame Relay and ATM networks, to user A, who also has a Frame Relay workstation. Neither is aware of the intervening ATM network, the necessary conversions having been carried out in the Interworking Functions (IWFs).   *IWF*

This technique of transparently carrying Frame Relay traffic over an ATM network is known as *Frame Relay Transport Over ATM*.

- A Frame Relay device should be able to communicate with an ATM device without the Frame Relay device being aware of the conversion – the ATM device performing the necessary conversion. This is illustrated in Figure 6.15 by user C wanting to talk to user B. User B has an ATM workstation and the Frame Relay traffic will arrive at, and be converted within, B's workstation.

- A Frame Relay device should be able to communicate with an ATM device without either device being aware of the conversion being carried out by an Interworking Function. This is also illustrated in Figure 6.15, although the IWFs are now very different. The distinction between this technique, known as Service (as distinct from Network) Interworking, is that B's workstation is unaware of the traffic having originated as Frame Relay.

There are two primary standards, both originating from the Frame Relay Forum, which cover these interactions. The first two types of interworking are known as *Network Interworking* and are defined in the standard FRF.5 (reference [70]). The third type, known as *Service Interworking*, is defined in standard FRF.8 (reference [71]). The three types of interworking are described separately in the subsections below. There are, however, a number of elements common to FRF.5 and FRF.8:

- Congestion. Frame Relay and ATM handle congestion in slightly different ways: as explained earlier in this book, ATM uses the concept of an Explicit Forward Congestion Indicator (EFCI) while Frame Relay frames carry Forward and Backward Explicit Congestion Notifications (FECNs *FECN* and BECNs). When a frame experiences congestion its FECN bit is set *BECN* and the destination then sets the BECN bits on frames it returns so as to alert the origin of the congestion.

  Both FRF.5 and FRF.8 call for the Frame Relay FECN and BECN bits to be stored in the SSCS header of the frame so that it can be restored at the destination. Both also give two modes of operating: one copying the Frame Relay FECN bit into the ATM EFCI field, and one using the EFCI field purely to represent congestion on the ATM network. The mode is chosen per connection by configuration.

- Cell-Loss Priority. Another key field within ATM when congestion occurs is the CLP carried within the ATM header. The equivalent field in the Frame Relay world is the Discard Eligibility Flag (DE) and FRF.5 and *DE* FRF.8 both give the mapping between these two fields. Again there are

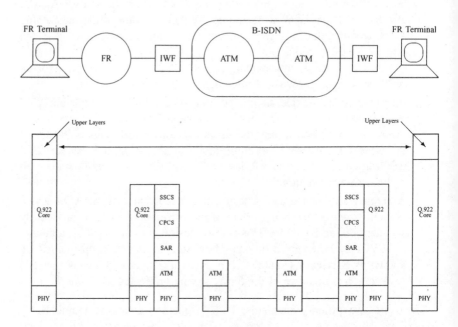

Fig. 6.16 *Frame Relay Transport Over ATM*

two modes, configurable by the operator, one of which copies between the CLP and the DE, the other of which sets the CLP to a constant value.

- Packet Reassembly. Although Frame Relay transports variable-length frames, it has a maximum length that it can handle. Frames larger than this have to be segmented for transmission across the Frame Relay network. Both FRF.5 and FRF.8 make it clear that reassembly of the frames before they are segmented into ATM cells is a function of the Frame Relay side of the IWF.

Having considered the points common to FRF.5 and FRF.8, the following subsections deal with their distinctions.

6.4.1 Frame Relay Transport Over ATM

This is the simplest of the three techniques to understand: the Frame Relay traffic is terminated by the Interworking Function (IWF) and passed across the ATM network where it is converted back to a Frame Relay interface. The workstations involved have no knowledge of the intervening ATM network.

Figure 6.16 illustrates the stacks involved; for traffic arriving from the left in the diagram, the IWF terminates the Frame Relay connection in accordance with standard Q.922 (reference [72]). The frames then descend through an ATM

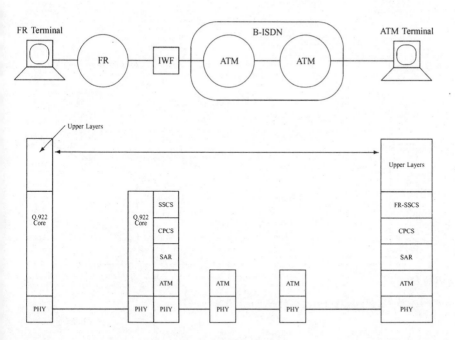

Fig. 6.17 *Frame Relay/ATM Network Interworking*

stack, using a special Frame Relay Service-Specific Convergence Sublayer (FR-SSCS). The FR-SSCS strips some fields from the Frame Relay frame (including *FR-SSCS* the cyclic redundancy check and flags) and removes the extra zero bits inserted to prevent a run of one-bits being mistaken for a flag (see glossary entry on bit stuffing) and passes the resulting frame to a standard AAL-5 Common Part Convergence Sublayer (CPCS) as defined in I.363 (reference [17]).   *CPCS*

The FR-SSCS can also provide multiplexing of several Frame Relay connections into one ATM Virtual Channel and can handle congestion (by mapping between the ATM and Frame Relay discard techniques).

6.4.2   Frame Relay/ATM Network Interworking

The interworking of a Frame Relay and an ATM workstation, where the conversion is carried out in the ATM workstation, is also defined in FRF.5 (reference [70]). This is very similar to the Frame Relay Transport Over ATM, described above, except that the stack containing the FR-SSCS is now present in the ATM workstation rather than in an IWF. The technique is illustrated in Figure 6.17.

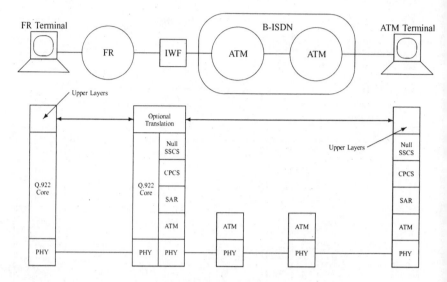

Fig. 6.18 *Frame Relay/ATM Service Interworking*

6.4.3   Frame Relay/ATM Service Interworking

The ideas of true Frame Relay/ATM Service Interworking (i.e. an ATM device communicating with a Frame Relay device without either being aware of the protocol conversion) is more complicated than Network Interworking and is defined in FRF.8 (reference [71]).

Again, the simplest way to view the protocol is through the interacting stacks, as shown in Figure 6.18.

The function performed by the IWF in FRF.8 is similar to the function performed in the ATM device in FRF.5 (see above), with one important difference: the Service-Specific Convergence Sublayer (SSCS) is now empty, its functions having been performed at the application layer within the IWF. The residual ('null') SSCS simply provides the correct interfaces to the Frame Relay protocol on one side and the AAL-5 Common Part Convergence Sublayer (CPCS) on the other.

Since the frames rise through the stacks at the IWF as high as the application layer, the IWF can be considered a Gateway.

In a similar manner to Network Interworking, frames travelling towards the ATM device have their Frame Relay header, Cyclic Redundancy Check and flags stripped and inserted zero bits removed before being segmented into AAL-5 cells. Cells travelling towards the Frame Relay device are reassembled into frames using the AAL-5 techniques for identifying frame boundaries, zero bits are inserted where necessary, and a Cyclic Redundancy Check and Frame Relay header are added.

FRF.8 closely defines the fields taken from the ATM or Frame Relay header to build the header for the other protocol.

7
ORGANISATIONS

Numerous companies and organisations are involved in ATM. This section gives contact points for the standards bodies and equipment manufacturers mentioned in this book.

7.1 Standards bodies

- International Telecommunications Union (ITU).   *ITU*

      International Telecommunications Union
      Information Services Department,
      Place des Nations, 1211 Geneva 20,
      Switzerland.
      Voice: +41 22 730-6666 or +41 22 730-5554
      Fax: +41 22 730 533
      Email: helpdesk@itu.ch
      Internet: http://www.itu.ch

- American National Standards Institute (ANSI).   *ANSI*

      New York Headquarters
      13th Floor, 11 West 42nd Street,
      New York, NY 10036, USA
      Voice: +1 212-642-4900
      Fax: +1 212-398-0023
      Internet: http://www.ansi.org

- European Telecommunications Standards Institute (ETSI).   *ETSI*

      ETSI Telecommunication,
      E-28040 Madrid, Spain.
      Voice: +34 1 336 7332
      Fax: +34 1 336 7333
      Internet: http://www.etsi.fr/home.htm

ATM Forum   • ATM Forum.

         Internet: `http://www.atmforum.com`
         Internet: `ftp://ftp.atmforum.com`

ATM25   • ATM25 Alliance.

         Voice: +1 408-383-1355
         Internet: `http://www.atm25.com/atm25.html`

IEEE   • Institution of Electrical and Electronic Engineers (IEEE).

         IEEE, 445 Hoes Lane, PO Box 1331,
         Piscataway, NJ 08855-1331, USA
         Voice: +1 908-981-0060
         Fax: +1 908-981-0027
         Email: `www-organizations@ieee.org`
         Internet: `http://www.ieee.org`

IETF   • Internet Engineering Taskforce (IETF).

         IETF Secretariat,
         c/o Corporation for National Research Initiative
         1895 Preston White Drive,
         Suite 100, Reston, VA 22091, USA
         Voice: +1 703-620-8990
         Fax: +1 703-620-0913
         Email: `ietf-info@ietf.org`
         Internet: `http://www.ietf.org`

7.2 Companies

Note that many companies are active in the ATM field. Only those companies explicitly mentioned in this book are included here.

Cisco   • CISCO.

         CISCO Systems Inc.
         170 W. Tasman Dr.
         San Jose, CA 95134-1706, USA
         Voice: +1408-526-4000
         Internet: `http://www.cisco.com`

IBM   • IBM.

         17 Skyline Drive,
         Hawthorne, NY 10532, USA
         Voice: +1914-784-3251
         Internet: `http://www.ibm.com`

- Ipsilon. *Ipsilon*

  232 E. Java Drive,
  Sunnyvale, CA 94089-1318, USA
  Voice: +1408-990-2000
  Internet: http://www.ipsilon.com

- Northern Telecom (NORTEL). *Nortel*

  2920 Matheson Blvd. East,
  Mississauga, Ontario, L4W 4M7, Canada
  Voice: +1905-566-3000
  Internet: http://www.nortel.com

- Toshiba. *Toshiba*

  1 Komukai Toshiba-cho,
  Saiwai-ku
  Kawasaki 210, Japan
  Voice: +8144-549-2238
  Internet: http://www.toshiba.com

7.3 Further ATM Information

One very useful entry point into the mass of documentation available on the
World-Wide Web is

    http://www.spp.umich.edu/telecom/telecom-info.html

This site contains pointers to all of the relevant standards bodies and many
of the equipment manufacturers. In addition it has some magnificent search
engines covering telecommunications research papers.

Another useful site is the Cell Relay Retreat which can be found at URL:

    http://cell-relay.indiana.edu/

The cell relay usenet conference, maintained by Indiana University, is useful
for asking questions relating to ATM and associated technologies. The conference subscribers are knowledgeable and have much experience in the area of
ATM. The news group URL is:

    news://comp.dcom.cell-relay

8
GLOSSARY

Glossaries are often of the type which tell you, correctly, that EPRCA stands for Enhanced Proportional Control Algorithm and then leaves you little the wiser. The authors hope that this glossary, used together with the index which follows, will do more. The index does not point to entries in this glossary. However, entries in the index are generally listed under a term's full title. Thus, for example, if the reader is puzzled by the acronym *MPOA*, he or she is expected to use this glossary to find that the acronym expands to *Multiprotocol Over ATM* and, if further information is required, to use the index to find references to *Multiprotocol Over ATM*.

Within the glossary, all ATM technical terms, acronyms and abbreviations are expanded and a page reference given. Where the term is a general telecommunications concept rather than being specific to ATM, a short tutorial is provided. The authors hope that this will make the glossary useful to readers with different levels of telecommunications experience.

It should be noted that the ATM Forum also provides an annotated glossary on its web page at

> http://www.atmforum.com/atmforum/glossary/glosspage.html

In ordering this glossary no distinction has been made between upper- and lower-case letters but digits have been put before letters. Punctuation marks such as hyphens have been ignored for the purposes of ordering.

0xAA. Hexadecimal (base 16) numbers in this book are written in this form (as in the C language). Thus 0xAA has the binary value 10101010 and denary value 170.

AAL. ATM Adaptation Layer. The AAL interfaces between the user and the ATM network and segments user data into ATM cells, reconstituting those cells into user data at the destination. See page 15. Specific interfaces to the AAL are defined on pages 40 (AAL-1), 44 (AAL-2) and 46 (AAL-3/4 and AAL-5).

AAL-CU. AAL-2 in its original form was never completed. A new adaptation layer has been proposed to replace it that incorporates multiple packets in a cell. This AAL has also been called AAL-6 and AAL-CU. See pages 16 and 44.

ABR. Available Bit Rate. The type of ATM connection (see also CBR, UBR and VBR) which, instead of defining in advance the bandwidth that a customer can use, provides flow control from the network to limit the information flow from the customer's terminal to a rate which the network can accept. The bandwidth available to the customer is therefore subject to change as the network congestion changes. This technique allows a network provider to exploit otherwise unused bandwidth on the network. See page 5.

ADSL. Asymmetrical Digital Subscriber Loop. The term *loop* is used within telecommunications circles to mean the cabling between the exchange site and a subscriber's premises. Since a large part of a carrier's investment is in the subscriber loops, a great deal of effort has been expended in finding ways to make better use of it. The twisted-pair cable of the loop was originally designed to carry analogue telephony signals but various techniques have been proposed to send digital signals with a greater bandwidth down it, one of these being ADSL. *Asymmetrical* implies that the speed of transmission is not equal in both directions: the downstream bandwidth (i.e. towards the subscriber) is greater than the upstream bandwidth (i.e. from the subscriber). This reflects the normal use of a link by a subscriber; Internet access, for example, requires very little upstream bandwidth (the odd keypress or button click) and a large downstream bandwidth (for the transmission of colour graphical pictures).

Digital Subscriber Loop technology is increasing in bandwidth at a breathless rate and various technologies (including ADSL and VDSL) are popularly combined under the generic form xDSL.

AFI. Authority and Format Identifier. The first octet of a B-ISDN address which identifies the type of address that follows. See page 67.

AIS. Alarm Indication Signal. A signal transmitted back along a connection to indicate that a path failure has been detected further downstream. See page 76.

Al. Alignment. An artificial field inserted to ensure that a following field is correctly aligned (e.g. to an octet boundary). See page 53.

ANSI. American National Standards Institute. See page 125.

AREQUIPA. Application REQuested IP over ATM. A proposal from the Swiss Federal Institute of Technology for a protocol to be used by IP applications to set up the IP connection over an ATM network. See RFC2170 (reference [45]) and page 91.

ARIS. Aggregate Route-Based IP Switching. See the glossary entry for *Label Switching* and page 115.

ARP. Address Resolution Protocol. A protocol used within IP networks when one terminal needs to find the physical address of another, given only its higher-level name. Effectively, the terminal needing to know the physical address transmits a message of the type 'Who knows the address of *name*?'

Since IP messages are broadcast, the terminal with the required name will hear the message and respond. See pages 88 and 98.

ARQ. Automatic Repeat Request. See page 60.

ATM. Asynchronous Transfer Mode. The low-level protocol used by broadband ISDN to transfer information from source to destination, the information being broken into 48-byte payloads in 53-byte cells. See page 2.

ATM25. An alliance of companies dedicated to bringing ATM at 25.6 Mb/s to workstations on the desktop. As there are still few desktop applications that make use of ATM's Quality of Service guarantees (which are largely what distinguishes it in a desktop environment from conventional LAN technologies), and even fewer ATM switches which really support the QoS guarantees, there is a belief that this technology is now a solution in search of a problem. See the information on the ATM25 alliance on page 126.

ATMARP. A protocol used instead of IP's ARP mechanism where the underlying medium is ATM rather than a broadcasting medium such as an Ethernet. See page 98.

ATM Forum. An organisation formed in November 1991 by parties interested in the emerging ATM system who felt that the normal speed of standards definition would be too slow for ATM. The Forum now has over 600 corporate members and sponsors a Technical Committee which meets every 2 months to revise and extend the base of ATM standards. The ATM Forum is not the only body producing standards for ATM and, although efforts are made to keep the standards in line, significant differences (particularly in the area of signalling) do exist between ATM Forum standards and those of the more traditional standardisation bodies.

ATMM-SAP. ATM Management Service Access Point. In addition to the n-SAP service entry points defined in the OSI model, ATM defines access points for the management (OAM) system. These are known as ATMM-SAPs. See page 27.

AUU. ATM-layer-user to ATM-layer-user parameter in an AAL-5 primitive to the AAL. See page 55.

Bandwidth. This is a term which has a precise meaning amongst communication specialists and a looser meaning amongst others. It is a measure of the amount of information which needs to be sent down a transmission line to carry a signal. For instance, conventionally coded telephone speech requires 64 kb/s of bandwidth. By compressing the speech this can be reduced to 32 kb/s, 16 kb/s or even 8 kb/s, at the price of increased distortion. A video picture requires many more bits per second to be sent down the line: a figure of 6 Mb/s for a compressed video of VCR quality often being quoted.

BASize. Buffer Allocation Size. A 2-octet value passed in the overhead of an AAL-3/4 cell to allow the receiver to allocate memory for the entire reassembled frame when the first cell is received. This means that it is unnecessary to allocate maximum-sized buffers at the receiver to ensure that the frame can be reassembled. See page 54.

BCD. Binary-Coded Decimal. A technique whereby a number is stored in a computer in decimal form using 4 bits for each digit. Thus the decimal number 1997 would be stored in 2 bytes as 0x1997. This method of storage is not as economical with space as pure binary (0x1997 requires 16 bits whereas 1997 in binary requires only 11 bits) but it is much easier to decode to a denary number.

BECN. Backward Explicit Congestion Notification. A single-bit field within the header of a Frame Relay frame indicating that congestion was detected somewhere along the opposite direction path, indicated by FECN being set on incoming frames. See page 119 and the description of FECN in this glossary.

B-ISDN. Broadband ISDN. The communications protocols used to transfer information from various services (speech, video and data) across the same network. See page 1.

B-ISUP. Broadband ISDN User Part. An extension into the broadband environment of the interswitch ISDN User Part (ISUP) signalling of Signalling System Number 7 (SS7). The B-ISUP standard is an ITU standard and is not in total accord with the equivalent ATM Forum standard, B-ICI: another source of annoyance for equipment manufacturers trying to sell into both the European (ITU) and North American (ATM Forum) markets. See page 64.

Bit stuffing. In some protocols, such as HDLC, frames are terminated by a specific bit pattern known as a flag. This bit pattern must not occur in the data being transferred, otherwise the receiver will be fooled into believing that the frame has ended. A flag is normally a consecutive sequence of binary ones and, to avoid data being interpreted as a flag, the transmitter stuffs extra binary zero bits into the data when it detects a run of one bits. These are removed by the receiver before the data are transferred to the final destination. These extra zero bits are of importance when Frame Relay frames (which have bit stuffing) are converted from or to ATM cells (which do not). See page 121.

BLINKBLT. Broadband Link-Layer Block Transfer, a proposal for an improved ATM Adaptation Layer for data transmission. See page 60.

BOM. Beginning Of Message. COM and EOM refer to Continuation Of Message and End Of Message respectively. When a frame is segmented into cells for transmission, unless it can be contained in a single cell (a Single Segment Message (SSM)), it is transmitted as a BOM, a number of COMs and an EOM. See page 51.

Bridge. A device which transfers messages at the Link Layer (layer 2) between two LAN segments. See page 93.

BSVCI. Broadcast Signalling Virtual Channel Identifier. See page 69.

Btag. Beginning Tag is used in AAL-3/4 to mark the start of a CPCS-PDU. The tag must have the same value as the End Tag to indicate error-free transmission. See page 54.

BUS. Broadcast/Unknown Server. A server on an Emulated LAN (ELAN) which, by establishing a point-to-multipoint connection across the ATM network with all clients on the Emulated LAN, serves as a focal point for broadcasts. Frames sent to it (on point-to-point ATM connections) are retransmitted to all clients on the Emulated LAN, thus simulating the broadcast of a conventional LAN. See page 102.

CAC. Connection Admission Control. This is one of the most open areas of the ATM development at the moment. It refers to an ATM switch making the decision whether or not to accept a new call with particular Quality of Service guarantees. Accepting a call and then finding that the Quality of Service cannot be maintained might lead to complaints from subscribers; not accepting a call when the call could have been handled leads to loss of revenue for the network provider. See page 71.

Cache. A term referring to a fast, short-term memory.

Carrier. A term used generically in this book to refer to companies selling telecommunications bandwidth to subscribers. In many countries these companies were originally coupled with the Post Office and the term PTT (Post, Telephone and Telegraph) is still common in many European countries. The precise definition of a carrier is left vague here since it differs from country to country, but large national PTTs such as BT, France Telecom, Deutsche Telekom and Telefonica would all count as carriers.

CBR. Constant Bit Rate. The type of ATM connection (see also VBR, UBR and ABR) which provides the customer with a 'pipe' into which data can be poured at a constant rate. See page 4.

CCITT. Comité consultatif international télégraphique et téléphonique. Until the beginning of March 1993, all of the references in this book to the ITU would have read CCITT. The CCITT was the committee within the ITU which was responsible for telecommunications standards until that time.

Cell. Used within this document to mean a 53-octet ATM cell. In contrast to a 'frame', a cell is of a fixed length. See page 1.

Cell Relay. Another term for ATM, based on the analogy with Frame Relay.

Cell Tax. The rather negative term used to describe the approximately 10% overhead carried on each ATM cell. See page 8.

Classical IP. A strange term used to describe IP running over ATM in accordance with RFC1577 (reference [52]). The term is explained within that reference as follows: *The 'classical' model here refers to the treatment of the ATM host adapter as a networking interface to the IP protocol stack operating in a LAN-based paradigm.* See page 96 and section 6.3.1.

CLNAP. Connectionless Network Access Protocol. See page 57.

CLNIP. Connectionless Network Interface Protocol. See page 57.

CLP. Cell Loss Priority. This is a one-bit field, carried in the 5-octet header of an ATM cell to indicate the relative priority of the cell. Setting this field to 1 is an indication to the network that it can be discarded first if the network becomes congested. The originating user may set this bit or it may be set by the source switch (if, for example, the user is exceeding its guaranteed bandwidth). See pages 27 and 30.

CLSF. Connectionless Service Function. See page 57.

COM. See BOM.

CPCS. Common Part Convergence Sublayer. See pages 46 and 121.

CPE. Customer Premises Equipment. Terminal installed on the subscribers' premises (and normally belonging to the subscriber rather than to the network operator).

CPI. Common Part Indicator. A 1-octet field in the CPCS-PDU for AAL-3/4 that determines the BASize and Length field semantics. Currently the only defined encoding is of 0x00 which means that the BASize field dictates the buffer allocation in octets and the length of the CPCS-PDU payload is measured in octets. See page 46 and following pages.

CRC. Cyclic Redundancy Check. A cyclic redundancy check is a value transmitted with a message to enable errors in the transmission to be detected and, in some cases, corrected at the receiving end. As with most things in telecommunications, the CRC is a compromise: sending the additional information requires additional bandwidth on the link and it may happen that random errors will occur in such a way that the CRC is still correct: the error therefore not being detected.

The CRC is generated by treating the binary digits of the message as the coefficients of a polynomial, dividing that polynomial by a standard polynomial (e.g. $x^8 + x^2 + x + 1$) and then using the remainder. See page 35.

Credit-Based Flow Control. A technique which for a while was seen as a counter-proposal to Rate-Based Flow Control for controlling the amount of traffic a terminal could put into the network when using ABR. See page 5.

CS. Convergence Sublayer. See page 30.

CSI. Convergence Sublayer Indication. See page 41.

CSR. Cell-Switch (or *switched*) Router. A proposal by Toshiba for the efficient transfer of IP packets over an ATM network. See the glossary entry for *Label Switching*, page 112 and RFC2098 (reference [66]).

Cut-Through Connection. A term used within a LAN Emulation or MPOA environment to refer to a direct ATM connection built between two end points to avoid the need for sending packets to a Broadcast Server (BUS) in the case of LAN Emulation or to an MPOA Server in the case of MPOA. See page 103.

DCC. Data Country Code. See page 67.

DE. Discard Eligibility. A flag carried in the header of a Frame Relay frame to specify whether the frame should be discarded if congestion were to occur. This flag is equivalent to the CLP flag in the header of ATM cells. See page 119.

Demand Connection. See *Switched Connection*.

Denary. The denary number system is that using base ten (i.e. the normal numbers used by the person on the street). Sometimes this number system is called the *decimal* system.

Digitisation. The act of turning an analogue signal (such as the electrical signal from a microphone) into a stream of numbers: each number being termed a *sample*. A basic theorem of telecommunications (after Nyquist) states that if samples of an analogue signal are taken at a rate at least twice as fast as the highest frequency component in the signal, then the original analogue signal can be exactly reproduced from the samples. Analogue telephony is normally filtered to the range 300–3200 Hz and then sampled at 8000 eight-bit samples per second (i.e. once every 125 μs), and, since this is more than twice the highest frequency present in the signal, the filtered voice can be reconstructed exactly from the samples at the receiver.

DLCI. Data Link Control Identifier. The identifier of a subscriber in a Frame Relay network.

E3. A link speed: see the glossary entry for *plesiochronous*.

EFCI. Explicit Forward Congestion Indicator. A bit in the header of an ATM cell set by switches through which the cell passes if those switches are congested. See page 28.

ELAN. Emulated LAN: a local area network emulated, using LANE protocols, by an ATM network. See page 100.

Encapsulation. The carrying of traffic encoded in one protocol by another protocol without the data being decoded. See page 87.

Enterprise Network. A network owned (and normally operated) by a company or university for its own internal use. See page 17.

EOM. See BOM.

EPD. Early Packet Discard. The ATM technique (applicable to AAL-5) of trying to discard all of the cells of one packet rather than one cell from each of many packets when congestion occurs. See page 95.

EPRCA. Enhanced Proportional Control Algorithm.

ESI. End-System Identifier. See page 68.

Etag. End Tag is used in AAL-3/4 to mark the end of a CPCS-PDU. The tag must have the same value as the Beginning Tag to indicate error-free transmission. See page 54.

Ethernet. A local area network (LAN) where several workstations are physically connected to the same cable and can therefore hear all transmissions on that cable. When a workstation needs to transmit to another workstation on the cable, it waits until the cable is free and then transmits its message. If a collision occurs because two workstations transmit at the same time, then both stop transmitting and wait a random time before trying again. All of the workstations on the Ethernet receive the message but all except the intended destination discard it. Ethernets have traditionally operated the shared cable at 10 Mb/s, but 100 Mb/s and even 1 Gb/s products are now becoming available. See also *Token Ring*.

ETSI. European Telecommunications Standards Institute. See page 125.

Extranet. Whereas an Intranet is an Internet-like network within one company (sharing forms, documents, etc. within the company) an Extranet is an extension of this to include the networks of the company's suppliers and customers. See page 89.

F1–F5. These represent five different levels at which a Virtual Channel needs to be monitored: F1 refers to the lowest physical level and F5 to the end-to-end Virtual Channel itself. See page 74.

Fabric. A term for the central core of an ATM Switch: the element (often an integrated circuit) which actually performs the switching. See page 10.

FANP. Flow Attribute Notification Protocol. A term devised by Toshiba for their protocol used between neighbouring nodes to establish cut-through paths for ATM cells carrying IP traffic. See RFC2129 (reference [60]) and page 114.

FCAPS. An acronym made from the initial letters of Fault Management, Configuration Management, Accounting Management, Performance Management and Security Management. These are the areas which a Network Management System is expected to cover. See page 80.

FCS. Frame Check Sum. A value transmitted with a frame so that an error in transmission can be detected.

FEC. Forward Error Checking. A code used to detect and correct corrupted octets. See page 43.

FECN. Forward Explicit Congestion Notification. A single-bit field within the header of a Frame Relay frame indicating that congestion has been detected somewhere along the path that the frame has taken. See page 119 and the description of BECN in this glossary.

FERF. Far End Receive Failure. A type of OAM cell which is sent 'upstream' (i.e. back towards the originator) by a switch when a failure has caused one of its incoming Virtual Channels or Virtual Paths to fail. See page 77.

FIB. Forwarding Information Base. A term used to describe the table held within a router, containing information about possible destinations and which interface to use to move one hop towards the destination. See page 109.

Firewall. A device or program which protects a private network (normally an IP-based LAN) from malicious or accidental access by unauthorised users. The following description is taken from reference [73]: *The security experts say that a firewall is a dedicated machine that checks every network packet passing through and which either drops or rejects certain packets based on rules set by the system administrator. ...It is now common practice to call anything that filters network packets a firewall.* See page 111.

Flat Addressing. See *Hierarchical Addressing*.

Flow Control. The technique whereby a device (or network) protects itself from overload by signalling back to tell the source to stop transmitting. This technique is as old as computing itself: early terminals used to use protocols whereby they sent X-ON and X-OFF characters back to the source to stop or restart transmission, normally by halting or restarting the paper-tape reader. Within ATM networks two techniques are used: EFCI and Rate-Based techniques. See page 28.

FR. See *Frame Relay*.

Frame. The term 'frame' is used in this book to mean a chunk of information to be transmitted across the network. Unlike a 'cell', a frame may be of any length up to some predetermined maximum depending on the protocol being used.

Frame Relay. A technique for transferring data across a wide area network. The technique grew out of X.25 Packet Switching when it became clear that, as links became of better quality, the heavy X.25 protocol was no longer needed. For details of the interworking of ATM and Frame Relay see page 118.

FR-SSCS. Frame Relay Service-Specific Convergence Sublayer. A particular Service-Specific Convergence Sublayer (see page 46) designed for Frame Relay. See page 121.

ftp. File Transfer Protocol. A protocol which makes use of TCP and IP to transfer bulk data (typically files) from one machine to another. See page 94.

Gateway. An application program which interfaces between two networks, accepting messages from one and, as applicable, re-transmitting them onto the other. See page 93.

GFC. Generic Flow Control, part of the ATM cell header at the User–Network Interface. See page 28.

GSMP. Generic Switch Management Protocol. The protocol used in Ipsilon's IP Switching between the ATM Switch and the colocated IP router. In fact the protocol could have far wider application, being a general-purpose protocol for establishing and releasing ATM connections, adding and removing leaves from point-to-multipoint connections and generally managing an ATM Switch by collecting statistics, requesting configuration information, etc. See page 111 and RFC1987 – reference [65].

Handshaking. An exchange of information using some standard protocol. It is often used to describe the initial exchanges which take place when end devices first make contact with each other.

HEC. Header Error Control. A cyclic redundancy code that identifies errors in the header. See pages 30 and 33.

Hierarchical Addressing. In any telecommunications network there must be a means of identifying each connected terminal (telephone, workstation, etc.) uniquely. Two different techniques are commonly used:

- Within a small network it is often convenient to assign each terminal a unique number effectively at random. Terminals on an Ethernet, for example, use the 48-bit MAC addresses built in at the time of manufacture. There is no structure in this address, it is simply the next 48-bit number in sequence at the factory. This is an example of a non-hierarchical or flat addressing scheme.

- Within a large network, a flat addressing scheme becomes unwieldy and some form of structure has to be imposed. A telephone number is a good example: +1 613 747 8700 breaks down hierarchically into +1 = North America, 613 = Ottawa, 747 = Eastern Ottawa, 8700 = unique number within Eastern Ottawa. Another example of a system with a hierarchical address scheme is the Internet where an address consists of a network identity followed by a terminal (host) identity within that network. With the advent of Local Number Portability (i.e. the ability for a telephone subscriber to take a telephone number with him or her when he or she moves house), even the hierarchical structure of telephone numbers appears to be breaking down.

The advantage of a hierarchical address scheme in a large network is clear; if telephone numbers were distributed world-wide in a flat, random fashion and we were to make a call from Godmanchester to Eastern Ottawa then the exchange in Godmanchester would have to know about all telephone numbers in the world in order to decide in which direction to route the call. With the current, hierarchical, system, the exchange in Godmanchester simply has to recognise the international +1 prefix as being North America and then pass the remainder of the number, without interpretation, across the Atlantic.

HLPI. Higher-Layer Protocol Identifier. This field indicates the destination higher layer entity of a frame and is used in the Connectionless Network Access Protocol (CLNAP). See page 58.

HO-DSP. High-Order Domain-Specific Part: part of an NSAP address. See page 68.

ICD. International Code Designator: part of an NSAP address. See page 68.

IETF. Internet Engineering Task Force. This body, which co-ordinates the RFCxxxx standards, is described as follows on its web page:

The Internet Engineering Task Force is a loosely self-organized group of people who make technical and other contributions to the engineering and evolution of the Internet and its technologies. It is the principal body engaged in the development of new Internet standard specifications. Its mission includes:

- *Identifying, and proposing solutions to, pressing operational and technical problems in the Internet;*
- *Specifying the development or usage of protocols and the near-term architecture to solve such technical problems for the Internet;*
- *Making recommendations to the Internet Engineering Steering Group (IESG) regarding the standardization of protocols and protocol usage in the Internet;*
- *Facilitating technology transfer from the Internet Research Task Force (IRTF) to the wider Internet community; and*
- *Providing a forum for the exchange of information within the Internet community between vendors, users, researchers, agency contractors and network managers.*

The IETF meeting is not a conference, although there are technical presentations. The IETF is not a traditional standards organization, although many specifications are produced that become standards. The IETF is made up of volunteers who meet three times a year to fulfil the IETF mission.

There is no membership in the IETF. Anyone can register for and attend any meeting. The closest thing there is to being an IETF member is being on the IETF or working group mailing lists. This is where the best information about current IETF activities and focus can be found.

IFMP. Ipsilon Flow Management Protocol. This is the protocol used between routers in IP Switching (Ipsilon's proposed method of transporting IP traffic efficiently across an ATM network). See page 110 and RFC1953 – reference [59].

IISP. Interim (or, more recently, Integrated) Inter-Switch Signalling Protocol – the first attempt to produce a standard interface been two ATM switches in the same private network. For this reason it is sometimes known as P-NNI version 0. See page 18.

ILMI. Interim Local Management Interface or Integrated Layer Management Interface. See page 83.

inATMARP. A protocol used (within Classical IP over ATM) to inform the ATMARP server of the ATM address and IP address corresponding to a particular end station. See page 98.

InterNIC. Internet Network Information Centre. This is the organisation which allocates IP addresses to organisations. See page 93.

Intranet. A private Internet. This is a network offering the flexibility of the Internet (hypertext pages on-line, etc.) within a single company. Companies are finding this format increasingly useful for disseminating internal information, particularly as it can then be read equally easily by users with workstations, PCs or Macintosh computers. See page 89 and the glossary entry for Extranet.

IP. Internet Protocol. An internetwork layer protocol developed for data networks. Other common internetwork layer protocols, with which IP may be compared, are IPX (from Novell) and AppleTalk. The IP is described on page 92 and the means of efficiently carrying it over an ATM network is the subject of section 6.2.

IP Switching. See the glossary entry for *Label Switching* and page 110.

IPv4. Internet Protocol version 4. This is currently the most widely deployed version of the IP. See page 92.

IPv6. Internet Protocol version 6 (sometimes called IPng). This is the proposed version of the IP bringing with it advantages in address space, support for Qualities of Service, etc. See page 93.

IPng. Internet Protocol – next generation. This term is synonymous with IPv6.

IPX. Novell Internetwork Packet Exchange. This is an internetwork protocol developed by Novell for use in the Novell Netware family of products. It derives from the Xerox Network System and, with the increase in IP usage, is becoming increasingly marginalised.

ISDN. Integrated Services Digital Network. ISDN comes in three flavours: Basic Rate ISDN (BRA), Primary Rate ISDN (PRI) and B-ISDN:

- Basic Rate ISDN is an attempt to offer a mixture of computer and telephony services using the existing twisted-pair wiring to subscribers' homes. It supports three logical circuits to each subscriber: two 64 kb/s data circuits, called the B-channels, and one 16 kb/s signalling circuit, called the D-channel. For this reason, basic rate ISDN is termed 2B+D. Commonly one B channel would be used for telephony and the other for a computer connection to the Internet although this is not predefined and both B channels are often used for data, giving a 128 kb/s link. ISDN is sometimes considered a failure because it has never really become a popular service for domestic subscribers. This is due more to the lack of marketing than the technology: in Germany, where it was actively marketed by Deutsche Telekom, domestic penetration of ISDN is relatively high. In North America it was never actively marketed and domestic penetration is only now beginning as people look for access to the Internet at speeds higher than analogue modems can achieve.
- Primary rate ISDN is based on a similar technique but makes use of a 2 Mb/s link and offers 30B+D (i.e. 30 × 64 kb/s links and a single signalling channel).

- Broadband ISDN is the primary subject of this book.

ISO. International Standards Organisation. The ISO is an international agency, founded in 1946, to prepare standards in many areas (not just telecommunications). It is the standardisation body which prepared Open Systems Interconnection architecture. Traditionally the ISO has been involved with higher-level standards than the CCITT. See page 23.

ISR. Integrated Switch Router. The combination of ATM switch and IP Router proposed in IBM Corporation's ARIS protocol for the efficient transfer of IP packets across an ATM network. See page 115.

ITU. International Telecommunication Union. In its own words, the ITU is 'an international organisation within which governments and the private sector co-ordinate global telecommunication networks and services. Activities include the co-ordination, development, regulation and standardisation of telecommunications and the organisation of regional and world telecom events'. See page 20.

IWF. Interworking Function. This term is normally used to describe a device which converts from one protocol to another. Typically it will contain two stacks: incoming data will ascend through one stack and descend through the other for transmission. See Figure 6.15 and page 119 for an example.

Label Switching. This concept is explored in detail starting on page 106 but can be summarised as a technique whereby IP frames travelling as cells across an ATM network (or, indeed, other type of network) are *tagged* or *labelled* on entry to the network. This allows switches to forward the cells without having to reconstitute the frames and route according to the IP-level destination. Many proprietary techniques have been proposed (and, indeed, in many cases, implemented) including IP Switching from Ipsilon, CSR from Toshiba, Tag Switching from CISCO and ARIS from IBM. In an attempt to pull these disparate standards together, the IETF is working on a standard called MPLS.

LAN. Local Area Network. A network, normally high speed (typically 10 Mb/s or 100 Mb/s with 1 Gb/s now starting to be delivered), joining computers within a limited area (a floor of an office building, laboratory, etc.). A LAN can be contrasted with a metropolitan area network (MAN) or wide area network (WAN).

LANE. ATM Local Area Network (LAN) Emulation. The technique whereby an ATM network emulates a LAN. See page 99 and reference [53].

LANE Service. The rather strange collective name used for the three LANE Servers: the LECS, the BUS and the LES. See page 101.

Layer. The term *layer* is used in this book to refer to one quasi-horizontal slice of a communications stack. Thus the ISO OSI model is sometimes referred to as the '7-layer model' as it defines seven effectively independent protocols.

LEC. LAN Emulation Client. The program which runs in each ATM device connected to an Emulated LAN (ELAN) to make the higher layers believe that they are connected to an Ethernet or Token Ring LAN. See page 100.

LECS. LAN Emulation Configuration Server. The device which a LAN Emulation Client contacts when it is first invoked to find out the address of its LAN Emulation Server (LES). See page 102.

LES. LAN Emulation Server. The device which holds information (in particular addressing information) about all of the LAN Emulation Clients on an Emulated LAN. See page 102.

LI. Length Indication Field. See page 53.

LIS. Logical IP Subnetwork. A term used in RFC1577 (reference [52]) to refer to an IP subnetwork implemented using an ATM network. See page 98.

MAC. Medium Access Control Layer. Within a communications stack, this is the layer which lies between the link layer and the physical layer and controls actual access to the physical transmission medium. The term is normally used to refer to the layer on a LAN although it is actually much wider in meaning and ATM can be considered the MAC for B-ISDN. See pages 66 and 92.

MF. Mediation Function. Reasonably enough, this is the function carried out by a Mediation Device. A Mediation Device is, in principle, any device which converts one protocol into another but the term is commonly used within telecommunications to refer to the device which mediates between the network elements and the Management system. Mediation devices normally offer some form of standard interface 'upwards' to the Management system and a vendor-specific interface downwards to the network element. By using mediation devices, a network operator may control different vendors' equipment from one OAM system. See page 83.

MIB. Management Information Base. Network Management is based on an object-oriented abstraction of the actual hardware and software elements. Object Classes may represent real, physical entities such as switches, workstations, LANs, etc. or abstract concepts such as queues, routing algorithms or buffer pools. Each instance of these object classes is known as a *managed object*. Information about each managed object and its relationship to other objects is held in the MIB. Stylised commands may be performed by a suitably authorised operator on managed objects; typically they can be interrogated for their value, they can have their value changed, they can be reset to default values and they can be created and destroyed. See page 83 and Table 5.6.

MID. Multiplex Identification used within AAL-3/4. See page 51.

Modulo *n*. A term meaning that arithmetic should be carried out using the remainder when the 'real' answer is divided by n. Thus 5×4 modulo 3 is the remainder when 20 (the real answer) is divided by 3. Thus 5×4 modulo $3 = 2$. Within telecommunications, the term is normally used to refer to a counter which reaches some maximum value and then starts again at zero. A counter, for example, which is *modulo* 8 would take the values 0, 1, 2, 3, 4, 5, 6, 7, 0, 1, 2, . . .

MPC. MPOA Client (i.e. Multiprotocol Over ATM Client). A device which wishes to make use of MPOA to transfer IP packets to another MPOA Client. The MPC lies at the ingress point to the ATM network and, when a flow of traffic is detected, creates an ATM Virtual Circuit to the destination. This prevents the overhead of repeated processing in IP routers. See page 105.

MPEG. A technique for compressing moving video pictures for transmission across a digital data network. The MPEG-2 standard uses variable bandwidth dependent on the required picture quality and the type of pictures being transmitted. See reference [5].

MPLS. Multiprotocol Label Switching. See the glossary entry for *Label Switching* and page 117.

MPOA. Multiprotocol Over ATM. A technique which extends LAN Emulation (LANE) by allowing cut-through ATM VCs to be created across multiple IP subnets. See page 103 and reference [55].

MPS. MPOA Server (i.e. Multiprotocol Over ATM Server). A device which provides routing (as distinct from switching) functionality in an MPOA network. See page 105.

Multiplexing. This term refers to the merging together of several streams of traffic into one composite stream (rather like the traffic at the merging of two motorways). Normally this is done to carry multiple independent logical connections over one physical connection. The history of telecommunications is full of ingenious ways of performing this: by splitting the transmission medium by frequency (with signals transmitted simultaneously at different carrier frequencies) or in time (a burst from one input stream and then a burst from the second). Normally signals need to be *demultiplexed* again at the receiving end. Two types of multiplexing are common:

- In non-statistical multiplexing the common connection carrying the multiplexed signal is adequate for all of the input signals to be transmitting simultaneously. Using the motorway analogy, this would require the merged motorway to be able to handle the total traffic of all the merging motorways even if they were fully loaded.

- In statistical multiplexing the common physical connection is inadequate for all of the input signals – the system depends on the statistical behaviour of the inputs being such that not all sources will transmit simultaneously for a long period. In the motorway analogy this is equivalent to building the merged motorway too small to handle maximum load from the feeder motorways and then relying on the feeder motorways not all being full simultaneously. The analogy can be stretched a little further by postulating a large car park at the junction of the motorways to hold cars temporarily in case two feeder motorways both had a burst of traffic at the same time. The car park is a *buffer* to handle 'burstiness' on the input.

NBMA. Non-Broadcast Multi-Access. A term used to describe a non-broadcasting medium being used to support a protocol which would normally expect a broadcast medium. Thus, IP was designed to operate over a broadcasting LAN but when it is applied to an ATM network then the special constraints of the non-broadcasting medium must be considered.

NE. Network Element: normally a physical piece of equipment in the network being managed by the OAM system. See page 73.

NEF. Network Element Function. This is that function which represents a Network Element (i.e. an object within the network) to the Management system. See page 82.

NHRP. Next-Hop Routing Protocol. An IETF draft standard used by MPOA to remove the LANE restriction of a single Logical IP Subnet (LIS). In the NHRP model, a Next Hop Server is associated with each LIS. By communicating between themselves, the Next Hop Servers can provide devices in one LIS with the ATM addresses of devices in other LISs. By this means an end-to-end virtual circuit can be built by a device in one LIS to a device in another.

The NHRP proposed protocol has come under criticism regarding its exclusive use of point-to-point (rather than point-to-multipoint) Virtual Circuits and its scalability (i.e. whether it will operate effectively in a large network). See page 104 and reference [56].

NNI. This term refers either to a Private Network–Network Interface or to a Private Network–Node Interface. This is a distinction without a difference and either expansion refers to the routing protocol used between ATM switches within a private network. See pages 64 and 18.

Node. This is a vague term used in this book to mean a device in a network. A node might be a switch or end user's terminal depending on context.

N-SAP. Layer N Service Access Point. The interface in an OSI stack between layer N and layer N + 1. See page 26.

NSAP. Network Service Access Point. See page 67.

N-SDU. Layer N Service Data Unit. The item passed from layer N + 1 of a communications stack (such as the ISO OSI) to the layer beneath, layer N. The N-SDU contains user information encapsulated in an (N+1)-PDU. See page 26.

NT. Network Terminator. This is the demarcation point between a subscriber's terminal and the carrier's ATM network. See page 18.

OAM. Operation, Administration and Maintenance (sometimes abbreviated to OA&M) or Operation and Maintenance. This term encompasses the five *FCAPS* functions: Fault Management, Configuration Management, Accounting Management, Performance Management and Security Management within a network. This activity is normally performed through workstations connected to the network and is particularly challenging for carriers using equipment from different suppliers; the standards for OAM interfaces are not well defined and equipment manufacturers tend to provide only proprietary interfaces.

OAM has traditionally been of one of two models:

- Centralised database model. In this model some form of centralised device (normally a cluster of workstations) holds a database for all the devices in the network. When an operator reconfigures a device, this is written into the database and then sent to the device concerned. When an operator enquires about a particular parameter on a particular device then this request is satisfied from the database. This model is particularly applicable to networks containing devices without a great deal of local intelligence.
- Local database model. In this model, instead of being held at a central site, data are held on the individual devices themselves. This means that operator enquiries about the values of parameters have to be satisfied by exchanging information with the devices.

Each of these techniques has its own advantages and drawbacks. The centralised model allows fast operator access but can lead to co-ordination problems between the database and reality if contact between the central database and the device is lost. The localised model simplifies the data storage, especially in a heterogeneous network, but generally appears slower to the operator as contact has to be made with real devices even for the simplest enquiry.

Although OAM was omitted from the first issues of the ISO OSI layered model, it was added subsequently in the X.700 series of recommendations. The basic structure of OSI OAM, as brought forward for use in ATM OAM, is that of a set of managed objects within each network element. These managed objects are held in a Management Information Base (MIB) and are manipulated locally by a Management Agent. This agent receives instructions from the (normally centralised) Management Process and translates these instructions into local hardware commands. Similarly the agent translates alarms arising in the managed objects into notifications to send to the Management Process.

Refer to page 20 and to the whole of Chapter 5.

OC-3. One of the standard synchronous transmission standards used in the SONET (North American) system. Running at 155 Mb/s, it corresponds to STM-1 in the SDH system used elsewhere in the world. See page 32.

Octet. An 8-bit quantity. Indistinguishable from a byte to most humans and to all modern computers. This was not, however, always so as some early computers had non-8-bit bytes and so the term *octet* was coined to mean an 8-bit value.

OSF. This term, confusingly, has two meanings:

- The Offset Field used within an AAL-2 cell. See page 45.
- The Operations Systems Function within the OAM system: see page 83.

OSI. Open Standards Interconnection. A standard prepared by the ISO (hence the nice palindromic, and invertible, acronym ISO OSI) defining an architecture for the interconnection of equipment from different manufacturers. See page 23.

OSPF. Open Shortest-Path First. A routing protocol used between routers in IP networks. This algorithm has several advantages over the earlier RIP algorithms: it chooses a path through a network based on a user-defined measure (metric) associated with each link (which can be associated with parameters such as transmission cost or delay) rather than simply on the number of hops. See page 109 and RFC2178 – reference [74].

Packet. A term used in this book to be almost synonymous with frame. It differs from a cell in that all cells are the same length whereas packets are of differing lengths.

Payload. That part of a cell or frame which is carrying useful (as opposed to control) information.

PCI. Protocol Control Information. The additional information added by a layer in an ISO stack to each frame travelling downwards through it. This information is read and removed by the peer layer on the receiving computer. See page 26.

PDH. See *plesiochronous*.

PDU. Protocol Data Unit. A complete 'packet' including both data and header for the network layer under consideration.

Per-VC Queueing. A technique in ATM switch design whereby the cells waiting to be transmitted are queued with one queue for each Quality of Service for each outgoing Virtual Circuit. See page 14.

P format. Pointer format. In AAL-1 a pointer is used to delineate blocks of data. The payloads that carry a pointer are 'P format' payloads. See page 41.

Plesiochronous. Refer, if necessary, to the description of *transmission* given under the entry in this glossary for *telecommunications*. Transmission links between switches are traditionally based either on the Plesiochronous Digital Hierarchy (PDH) (see page 31) or the Synchronous Digital Hierarchy (SDH) (see pages 26 and 31).

Links in a plesiochronous network (normally a mesh) are effectively independent of each other and in most of the world are based on link speeds of 2 Mb/s (E1), 8 Mb/s and 34 Mb/s (E3). In North America the equivalent link speeds are 1.5 Mb/s (T1) and 45 Mb/s (T3).

Links in a synchronous network (normally a ring) are all synchronised

Speed	Plesiochronous		Synchronous	
	Outside N.America	North America	Outside N.America	North America
64 kb/s		DS-0		
1.5 Mb/s		DS-1 (T1)		
2 Mb/s	E1			
6.3 Mb/s		DS-2 (T2)		
34 Mb/s	E3			
45 Mb/s		DS-3 (T3)		
51.84 Mb/s				STS-1 (OC-1)
155.52 Mb/s			STM-1	STS-3 (OC-3)
274 Mb/s		DS-4 (T4)		
622.08 Mb/s			STM-4	STS-12 (OC-12)
2.5 Gb/s			STM-16	STS-48 (OC-48)
10 Gb/s			STM-64	STS-192 (OC-192)

Table 8.1 *Plesiochronous and synchronous transmission*

and link speeds are based on the SDH or SONET standards described in subsection 2.6.1.

The advantage of a synchronous network is clear when a transmission device needs to break up the traffic on an incoming link into transmissions onto two or more outgoing links. If the links are not synchronised then the device needs to buffer and realign the data for the outgoing path. If the links are synchronised then only a very small (and constant) amount of buffering is required.

For a list of the most common plesiochronous and synchronous transmission speeds see Table 8.1.

PRBS. Pseudorandom Binary Sequence. This is a sequence of bits generated by a specific algorithm, and therefore not strictly random, but having some of the characteristics of random bits. These are typically generated by a shift-register with feedback taken from one bit in the register exclusively-ORed with the bit being delivered. For an example of a PRBS in use, see page 36.

Primitive. A primitive is a message used in a protocol. Each message that can be passed across an interface is known as a primitive. For a frivolous example see the glossary entry for *Reference Model*.

Protocol. One of the best descriptions of protocols and their use in telecommunications networks is contained in *Network Protocols* by Andrew Tanenbaum included in reference [75].

Tanenbaum's paper describes a three-layer protocol by imagining two philosophers who want to communicate with each other. Philosopher 1 lives in an ivory tower in Kenya and speaks only Swahili. Philosopher 2 lives in a cave in India and speaks only Telugu. The philosophers represent layer 3 entities.

To convey his thoughts, Philosopher 1 passes his message, in Swahili, to his translator across the layer 3/layer 2 interface (the 2-SAP). The translator converts the message to a language he has agreed with his peer in India, let's say Dutch: this is his agreed level 2 protocol. He passes the resulting Dutch message to his engineer (across the 1-SAP). The physical mode of transmission may be anything from a carrier pigeon to a runner with cleft stick: this is determined only by the layer 1 protocol.

When the Indian engineer receives the message, he passes it to his translator for rendition from Dutch into Telugu. Finally the Indian translator passes the message, in Telugu, to his philosopher.

Two important points come out of this analogy:

- Each person (layer) thinks of his communication as being horizontal (to his peer in the other stack) while the communication is actually vertical (to the lower layer). Philosopher 1, for example, considers himself to be talking to Philosopher 2 (horizontally) whereas, in fact, he is talking to his translator (vertically).
- The protocols are completely independent. The philosophers can switch the subject at will, the translators can switch from Dutch to Spanish at will and the engineers can switch from carrier pigeon to runner at will, all without their choices affecting the other layers.

It is precisely for these reasons that telecommunications networks are defined in a layered manner: so that changes can be made to any layer without affecting the others.

P-NNI. See NNI.

Point-to-Point Connection. Three connection topologies are defined for ATM Virtual Channels (VCs) and Virtual Paths (VPs):

- Point-to-point. As its name suggests this type of connection links a single end-point to another single end point.
- Point-to-multipoint. This type of connection can be thought of as a tree with a single root and multiple leaves. In principle the root can transmit to all of the leaves and each leaf can transmit to the root. In practice, transmission from the leaves to the root may not be supported. Certainly leaves cannot communicate directly with each other.
- Multipoint-to-multipoint. This type of meshed connection is not currently supported directly within ATM but two main methods have been proposed:
    - each member of the multipoint group setting up a (unidirectional) point-to-multipoint connection to every other member of the group. This is a very complex technique as it involves a

large number (order N^2) of connections and some technique to inform all members of the group of all terminals leaving or joining.

    — nominating a central node (a *multicast server*) to which all messages are sent (point-to-point) and then redistributed (point-to-multipoint).

    This technique requires only order N connections and is therefore more scalable to large networks but has the problem of requiring a server which is both a bottleneck and a single point of failure.

It is important that ATM be able to support some type of multipoint connections (or some other form of multicast capability) as most traditional protocols, having been devised for broadcasting LANs, are based on this technique.

Point-to-multipoint connection. See *point-to-point connection.*

PSVCI. Point-to-point Signalling Virtual Channel Identifier. See page 69.

PTI. Payload Type Indicator: a three-bit field in the header of every ATM cell defining the type of information (OAM or user) that the cell is carrying. See page 28.

PTSP. PNNI Topology State Packet. See page 71.

PVC. Permanent Virtual Circuit (or 'Permanent Virtual Connection'): a permanent Frame Relay or ATM connection. See the entries for *switched connection* and *telecommunications* in this glossary.

QoS. Quality of Service. This is a set of measurable characteristics of a B-ISDN connection agreed between the network supplier and the subscriber in advance (and for which the subscriber will be charged). The following criteria are included in the definition of Quality of Service:

- Cell Error Ratio: percentage of received cells which are in error.
- Severely-Errored Cell Block Ratio: the percentage of received blocks (a pre-agreed number of cells) of cells with errors.
- Cell Loss Ratio: percentage of transmitted cells which do not arrive.
- Cell Misinsertion Rate: cells which are delivered to the destination although they did not originate from the source (probably due to a corrupted header).
- Cell Transfer Delay: the time taken to pass cells from source to destination.
- Mean Cell Transfer Delay: the cell transfer delay averaged over a pre-agreed number of cells.
- Cell Delay Variation: the variation in the delay of the cells.

To simplify the Qualities of Service, some generic groups have been defined:

- Unspecified, which is all that most equipment currently offers.

- Service Class A, designed for circuit emulation and constant bit rate video.
- Service Class B, designed for variable-rate audio and video.
- Service Class C, designed for connection-oriented data transfer.
- Service Class D, designed for connectionless data transfer.

Growth is expected in the number of service classes as this is one area where different equipment manufacturers can distinguish their products from those of their competitors.

Much of the complexity of ATM stems from the QoS guarantees and the complex CAC algorithms needed to ensure that they are met. The alternative approach, favoured primarily by those equipment manufacturers with roots in LAN networks, is to avoid all this complexity by simply providing vastly over-speed switches and links. See pages 20 and 58.

Rate-Based Flow Control. A technique which has triumphed over Credit-Based Flow Control for controlling the amount of traffic a terminal could put into the network when using ABR. See page 5.

Reed–Solomon Code. A scheme for encoding information which allows errors introduced during transmission to be corrected by the receiver. A Reed–Solomon (n, k) Code encodes m-bit symbols (typically $m = 8$ and symbols are octets) into blocks of $n = 2^m - 1$ symbols – i.e. $m(2^m - 1)$ bits. Thus a block of k symbols is expanded by the addition of $n - k$ redundant symbols. For more details of these types of encodings see any general-purpose communications textbook (e.g. reference [19]).

Reference Model. A reference model is an abstraction of real equipment, used to define the position of interfaces precisely. Thus a committee specifying a car might design a reference model to define the interfaces between the driver and the engine. This interface could then be termed the D interface and would contain functions like:

- driver primitive to cause the car to stop
- driver primitive to cause the car to turn left
- driver primitive to cause the car to reverse
- etc.

The idea would be that all car manufacturers would have to provide the D interface but that the specification would not define *how* the interface was to be implemented. So one manufacturer might fulfil the turning left requirement by providing a steering wheel, another might provide an anchor to be dropped on the left side of the car.

The ITU (or CCITT as it was then) defined a reference model for the narrowband ISDN interface and has extended this to the broadband ISDN interface as detailed on page 23.

Repeater. A device which, operating at the physical (electrical) level, amplifies signals received from one physical cable onto a second cable. See page 93.

RM. Resource Management Cell. A cell sent down a Virtual Channel and reflected back by the destination to collect congestion information about the intermediate devices carrying the Virtual Channel. See page 5.

Routing. The technique, used by IP, whereby packets received by a device known as a *Router* are passed up the stack to layer 3 (the Network Layer) where their destination is determined and the decision of how to pass the packet on to the next node in its journey is made. This technique is contrasted in section 6.2 with Switching and Bridging: similar processes but carried out at layer 2. See page 93.

S-AAL. Signalling ATM Adaptation Layer. See page 64.

SAR. Segmentation and Reassembly: the act of breaking a frame into cells or reassembling those cells back into a frame. The SAR sublayer forms part of the ATM Adaptation Layer. See pages 30 (data) and 65 (signalling).

SC. Sequence Count: part of the AAL-1 sequence number field. See page 41.

SDH. See *plesiochronous*.

SDT. Structured Data Transfer. See page 41.

SDU. See *N-SDU*.

Signalling. A transmission system has to carry two types of information: data and signalling. Taking a simple example, a telephone provides the exchange with the following signalling information:

- subscriber has lifted the handset
- subscriber has replaced the handset
- subscriber has dialled the digit *n*.

It is sometimes forgotten that signalling is also passed from the exchange to the telephone:

- start ringing
- stop ringing
- increment subscriber's private billing meter.

Unless the call is local, the local exchange will have to pass some or all of the dialled digits to another exchange. In the early days of telephony, transmission between the exchanges used the same circuits as for the voice signals. Nowadays, however, a separate channel carries signalling information from exchange to exchange: with the voice telephony network this signalling information is normally carried using the SS7 protocol.

Even with systems more sophisticated than basic telephony the same basic signalling information needs to be exchanged between the subscriber's terminal and the local connection point:

- I want to initiate a call (going off-hook).
- Here is the identity of the party with whom I wish to communicate (dialling).
- I wish to terminate the call (going on-hook).

Silence Suppression. A technique used in telephony to reduce the bandwidth required for transmission of speech. Typically at least half of a two-way telephone conversation consists of silence and, during this time, no information is transmitted. To prevent an unnatural silence at the receiver, the equipment at the destination will normally introduce a gentle background noise when no speech information is being received.

Single-Arm Router. This term is used to refer to an IP router which has only a single connection to a network: the packets to be routed arrive on the single physical connection, are routed onto a different virtual circuit, and return to the network on the same physical connection.

SITA. Switching IP Through ATM. A proposal by Telecom Finland for passing IP efficiently across an ATM network. See page 117.

SN. Sequence Number used in a cell header. See page 41 (AAL-1) and page 51 (AAL-3/4).

SNMP. Simple Network Management Protocol. A protocol defined initially for the OAM (control and provisioning) of LAN equipment where the OAM is relatively simple and the cost of a sophisticated workstation-based Network Management System could not be justified. SNMP is defined in RFC1157 (reference [43]) and has a simple set of primitives which operate on the Management Information Base (MIB):

- GET: used to retrieve information.
- GET-NEXT: used to 'walk' the MIB and retrieve information.
- SET: used to modify information.
- TRAP: used to report an alarm.

Although defined initially for LANs, SNMP is now used more extensively and, in particular, forms the basis of the Interim (or Integrated) Local Management Interface (ILMI) for ATM. See page 83.

SNP. Sequence Number Protection. See page 41.

Source Routing. A technique whereby the source of a frame or cell determines the entire route of the frame or cell through the network. This is in contrast to a routing system where each node simply knows enough of the network topology to forward the frame or cell to its next node. See page 71.

SONET. See *plesiochronous*.

SPID. Service Profile Identifier. See page 69.

SRTS. Synchronous Residual Time Stamp. Services which require synchronisation between source and destination can minimise the amount of data transferred for synchronisation by means of the SRTS Method. This method depends upon the source and destination having the same derived synchronised network clock. This is the case of SDH- and SONET-based systems where all devices are synchronised, though for non-synchronous packet-based physical layer transports another method, known as the *Adaptive Clock Method* is used. For more details see subsection 3.2.3.

SS7. Signalling System Number 7. This is a protocol used by voice switches to communicate signalling information between one another. See also the glossary entries for *signalling* and *B-ISUP*.

SSCF. Service-Specific Co-ordination Function. See page 65.

SSCOP. Service-Specific Connection-Oriented Protocol. See page 65.

SSCS. Service-Specific Convergence Sublayer. See page 46.

SSM. Single Segment Message. See the glossary entry for *BOM*.

SSP. Service-Specific Part. See page 65.

ST. Segment Type. This field indicates whether a message is beginning (BOM), ending (EOM), continuing (COM) or a Single Segment Message (SSM). See page 51.

Stack. The complete set of layers on a protocol, for example the ISO OSI, is known as a stack. The higher levels of a stack (above layer 2) are normally implemented in software and various companies have made programs implementing the ISO OSI stack available for sale. See page 23.

STM-1. This is the first level of the SDH hierarchy (155 Mb/s). See page 32 and Table 8.1.

SVC. Switched Virtual Channel (or 'Connection' or 'Circuit'): a temporary Frame Relay or ATM connection. See the entry for *switched connection* in this glossary.

Switched Connection. Telecommunications connections have traditionally been of one of two types: permanent or switched.

- Permanent connections are set up through the network's OAM system and remain in place, whether used or not, unless the network fails. If the network fails and recovers, then it automatically re-establishes the permanent connections. A leased circuit is a simple example of a permanent connection in the world of basic telephony. Billing for a permanent circuit is normally simple: at regular intervals the customer is billed an amount based on the reserved bandwidth of the connection, whether or not the connection has been used.

- Switched connections are established at the request of the subscriber, the address of the required partner being indicated by some form of signalling protocol. Normally, switched connections are not automatically re-established if the network fails and then recovers; it is left to the parties to signal a request to re-establish the connection. A traditional telephone call is an example of a switched connection. Billing for switched connections is more labour-intensive than for permanent connections; the time that the call is established has to be recorded along with the destination address. Another record is written when the call is disconnected and care must be taken to record the closing of the call if equipment fails. The amount of information so gathered presents a problem to the manufacturers of switches. Consider a switch which is handling short calls (such as connections for National Lottery machines or credit card authorisations). These switches are called on to handle the set up and clearing of several hundred calls per second. Assuming 200 calls per second and 50 bytes of information stored per event then this results in some 10 kb per second being stored. Typically this information is shipped twice each day to the OAM system, a transfer of some 400 Mbytes of data.

B-ISDN Switched Connections are also known as *Demand Connections.*

Synchrononous. See *plesiochronous.*

TA. Terminal Adaptor. A device, normally desk or wall mounted, which performs protocol conversion. Terminal Adaptors are seen by telecommunications manufacturers as a temporary means of introducing a new service and are used on terminal equipment without the correct interface hardware. Eventually, if demand is sufficient, the terminal manufacturers build an interface to the new service into their devices and then the need for terminal adaptors disappears.

Tag Switching. See the glossary entry for *Label Switching* and page 108.

TCP. Transmission Control Protocol. A transport-layer protocol often associated with IP (IP being the lower layer, TCP the higher). TCP is designed to offer a reliable service by using the unreliable IP. TCP is employed by such commonly used programs as ftp (File Transfer Protocol), telnet and rlogin. TCP and IP are often bound into the twin TCP/IP, but other protocols (such as UDP) do use IP and the twinning of TCP and IP is not necessary for either protocol. TCP is mentioned in several places in this book but, for details of its retransmission strategy, see page 94.

TDP. Tag Distribution Protocol. This is the protocol proposed by Cisco Systems Inc. for the distribution of Tags in the Tag Switching concept. See page 109 and reference [61].

Telecommunications. Telecommunications is the art of connecting together terminals (telephones, computers, etc.) at different locations and enabling data to flow between them.

There are three partners in a telecommunications link: the subscriber who has information to transmit, the subscriber to whom the information is to be sent and the carrier (or network provider) who supplies the network to make transmission possible.

In order to make a telecommunications system useful, it must be possible for the originating subscriber to indicate with which remote subscriber he or she wishes to communicate. For permanent connections this is done by informing the network operator who configures an appropriate circuit through the OAM system. A conventional telephone network has permanent connections known as leased lines. On the B-ISDN network such a connection is known as a Permanent Virtual Connection (PVC).

Circuits which are not permanently established are known as switched circuits and the user's terminal somehow needs to communicate the address of the other party to some form of switch. With a telephone, the user dials and the signalling is passed to the exchange either as a series of breaks in the line current (loop disconnect dialling) or as a series of musical chords (touch-tone dialling). With a B-ISDN terminal the address of the called party is passed across the User–Network Interface (UNI).

It is the responsibility of the switches in the network to build a connection between the two terminals. In the early days of telephony this connection was a copper wire end-to-end, now it is a virtual connection between computers.

Telecommunications divides into three basic disciplines:

- ACCESS. The access part of the network is concerned with getting signals from the subscriber's premises to the local exchange. Thus the access network includes all the copper wiring running underground or on poles and the junction boxes which one sees in the street. Recently the access network has been changing dramatically from a passive network of copper wires to a sophisticated network of optical fibre and electronic devices.

  A large part of the cost of any telecommunications system lies in its access network because of the expense of digging trenches to bury cables to individual houses.

- SWITCHING. Switches have traditionally been housed in exchange buildings and are responsible for the connection of a calling party to a called party. The switches interface to the access network and to each other.

  Switch technology is also undergoing a major change, particularly with the introduction of ATM technology, and large monolithic switches are probably a thing of the past as their component parts can now be distributed around the outside of an ATM network and can communicate with each other through that network.

- TRANSMISSION. Transmission is responsible for the high-speed links between the switches. In the last 15 years the industry has moved from plesiochronous links to synchronous rings and link speeds have increased dramatically.

Telnet. A protocol which makes use of TCP (and thereby IP) to allow a user on one workstation to log into another workstation across a network.

TIB. Tag Information Base. A term used within Tag Switching to refer to the forwarding table: the table which links the tags on incoming cells with the interface on which the cells are to be transmitted. See page 108 and Figure 6.11.

TMN. Telecommunications Management Network. This is a set of standard protocols used by OAM systems to control and monitor network devices. See page 80 and subsection 5.5.2.

Token Ring. A type of local area network (LAN) which avoids the problem of collisions inherent in the Ethernet by connecting the workstations in a logical or physical ring and passing a token around the ring. A workstation may only transmit when it has the token.

Topology. A term used in this book to characterise the interconnexion of nodes (switches) in a network. Thus nodes may be connected in stars, in rings or be fully intermeshed.

TTL. Time-to-Live. This is a field used in the header of an IP packet to prevent it wandering around a network for ever if routing tables contain a loop. The TTL field is decremented at each router that the IP packet passes. When it reaches zero the packet is discarded.

UBR. Unspecified Bit Rate. The type of ATM connection (see also VBR, ABR and CBR) which provides the user with no guarantees at all about bandwidth, cell loss ratio or cell delay. This level of Quality of Service was introduced because it was all that the first generation of ATM switches could actually offer. See page 5.

UBR+. A semi-unofficial term for UBR with Early Packet Discard enabled. See page 95.

UDP. User Datagram Protocol. A transport layer protocol which uses the IP internetwork layer protocol to deliver packets. UDP, unlike TCP, is not connection-oriented and does not offer guaranteed delivery. See page 92.

UME. UNI Management Entity. This is a concept used in the Interim Local Management Interface (ILMI). A UME is associated with each UNI in the network and exchanges information with the UMEs at the other end of the UNI link. See page 84.

UNI. User–Network Interface. See page 18.

URL. Uniform Resource Locator. This term, which is defined in RFC1738 (reference [76]), is used for the string which specifies the location and format of information on the Internet. The string has the general form:

<scheme>:<scheme-specific-part>

where the defined <scheme>s include ftp (File Transfer Protocol), http (Hypertext Transfer Protocol), gopher (Gopher Protocol) and telnet (Interactive Sessions). The <scheme-specific-part> has the general form:

//<user>:<password>@<host>:<port>/<url-path>

RFC1738 also defines the standard format to be used to quote a URL.

VBR. Variable Bit Rate. The type of ATM connection (see also UBR, ABR and CBR) which provides users with a 'pipe' into which they can pour information at varying bit rates. The peak bit rate and the allowed 'burstiness' of the traffic will have been prearranged between the network provider and the user. See page 4.

VBRnrt. Variable Bit Rate – non-real time. After the term *VBR* had been defined, it was realised that variable bit rate data can assume two forms: one where the data rate is variable and the data being carried are time sensitive (as in compressed video) and one where, in spite of the variable data rate, there is no time sensitivity (such as computer-to-computer data transfer). These two types of VBR were called VBRrt and VBRnrt respectively. See page 4.

VBRrt. Variable Bit Rate – real time. See the glossary definition of VBRnrt and page 4.

VC. Virtual Channel. A unidirectional 'pipe' for carrying ATM cells. The VC has a distinct identification (Virtual Channel Identifier, VCI) carried by all cells on the VC. See also VP and page 28. When several VCs are joined end-to-end to form a longer 'pipe', this is known as a Virtual Channel Connection (VCC): see page 9. Note that, although VC strictly stands for Virtual Channel, almost everyone in the industry believes that it stands for Virtual Circuit and uses Virtual Channel and Virtual Circuit interchangeably.

VC-4. Virtual Container – Level 4. Note that this term has nothing to do with Virtual Channels or Virtual Circuits. A VC-4 is the basic payload in an SDH system. See page 32 and Figure 2.7.

VCC. Virtual Channel Connection. See VC.

VCI. See VC.

VCID. Virtual Connection IDentifier. A term used by Toshiba within their FANP protocol (see RFC2129 (reference [60]) and page 114) to identify an ATM link between two Cell-Switch Routers. Unlike a normal ATM VC identifier, the VCID is the same at both ends of the link.

VC Merge. A concept whereby two Virtual Channels arriving at an ATM switch with the same destination are merged into one Virtual Channel. This procedure is more complicated than it sounds because the switch performing the merge must be aware of the type of ATM cells being carried; with some ATM Adaptation Layers (e.g. AAL-5) it is essential that the cells relating to a particular frame be delivered in order and without intervening cells from other packets. Thus, a switch performing VC Merging must be able to recognise and store AAL-5 cells to allow complete packets to be transmitted on the common path. At the time of writing very few ATM switches support VC Merging.

VP. Virtual Path. A group of Virtual Channels having a common origin and destination (see page 9). A VP is identified by a Virtual Path Identifier (VPI). See page 28. When several VPs are joined end-to-end the resulting connection is known as a Virtual Path Connection (VPC). See page 9.

VPC. Virtual Path Connection. See VP.

VPI. Virtual Path Identifier. See VP. This acronym is also used (although not in this book) to mean Virtual Private Intranet: a network of real or virtual routers, managed by a carrier on behalf of a company, providing the functions of an Intranet to the company.

WAN. Wide area network. A network which, in contrast to local and metropolitan area networks (LANs and MANs), extends over a long distance. A WAN might, for example, be used to connect two LANs in different towns. See page 89.

Well-known. This term was (we believe) first used in the TCP/IP protocol definition to mean predefined (and non-modifiable) identities for terminations. For example, within the TCP/IP protocol, a telnet server is always on TCP port 23. This is a *well-known* port number for the telnet server. The term has been carried forward into ATM where the UNI signalling information, for example, is carried in VPI = 0, VCI = 5. This is again a *well-known* address.

WSF. Workstation Function. See page 83.

X.25. A standard describing the manner in which end-customer devices may be connected to a packet-switched network (reference [1]). Although X.25 actually specifies the interconnection protocols, the term is actually used more loosely to describe the network itself: people talk of *the X.25 network*. In many ways X.25 can be considered the precursor of Frame Relay and ATM.

References

1. ITU-T X.25: *Interface between Data Terminal Equipment (DTE) and Data Circuit-terminating Equipment (DCE) for Terminals Operating in the Packet Mode and Connected to Public Data Networks by Dedicated Circuits*, October 1996
2. J.M.Simmons and R.G.Gallager: *Design of Error Detection Scheme for Class C Service in ATM*, IEEE/ACM Transactions on Networking, Volume 2, Number 1, February 1994
3. S.Dravida and R.Damodaram: *Error Detection and Correction Options for Data Services in B-ISDN*, IEEE Journal on Selected areas in Communication, Vol. 9, No. 9, Dec. 1991
4. R.Händel, M.N.Huber and S.Schröder: *ATM Networks Concepts, Protocols, Applications*, Addison-Wesley Publishing Company, Second Edition, 1994
5. ISO/IEC-13818: *Information Technology–Generic Coding of Moving Pictures and Associated Audio Information*, 1996
6. J.Beran: *Statistics for Long-Memory Processes,* (Number 61 in the series *Monographs on Statistics and Applied Probability*), Chapman & Hall, 1994
7. H.E.Hurst: *Long-term Storage Capacity of Reservoirs*, Trans. Am. Soc. Civil Engineers, 116, 1951
8. ITU-T I.150: *B-ISDN Asynchronous Transfer Mode Functional Characteristics*, Helsinki, March 1993
9. ITU-T I.113: *Vocabulary of Terms for Broadband Aspects of ISDN*, Geneva, April 1993
10. ITU-T I.362: *B-ISDN ATM Adaptation Layer (AAL) Functional Description*, Helsinki, March 1993
11. ATM Forum: *ATM User–Network Interface (UNI) Specification, Version 3.1.* Note that version 4.0 of this standard is now also available.
12. ITU-T I.610: *B-ISDN Operation and Maintenance Principles and Functions*, Helsinki, March 1993
13. T.Knowles, J.Larmouth, K.G.Knightson: *Standards for Open Systems Interconnection*, BSP Professional Books, 1987 (ISBN 0-632-01868-2)
14. C.Newall, M.Matthews, F.Mazda, P.Newcombe, S.Rogers, C.Lilley and P.Cherng: *Synchronous Transmission Systems*, Northern Telecom Europe Limited, 1993
15. ITU-T I.361: *B-ISDN ATM Layer Specification*, Helsinki, March 1993
16. ITU-T I.371: *Traffic Control and Congestion Control in B-ISDN*, Helsinki, March 1993
17. ITU-T I.363: *B-ISDN ATM Adaptation Layer (AAL) Specification*, Helsinki, March 1993
18. ITU-T I.432: *B-ISDN User–Network Interface and Physical Layer Specification*, Helsinki, March 1993
19. S.Haykin: *Communication Systems*, John Wiley and Sons, 1994

20. ITU-T I.311: *B-ISDN General Network Aspects*, August 1996
21. ITU-T I.363.2: *B-ISDN ATM Adaptation Layer Type 2 Specification*, Toronto, September 1997
22. ITU-T I.211: *B-ISDN Service Aspects*, Helsinki, March 1993
23. ITU-T I.364: *Support of Broadband Connectionless Data Service on B-ISDN*, Helsinki, March 1993
24. ITU-T E.164: *Numbering Plan for the ISDN Era*, 1991
25. R.Iwase and H.Obara: *A Bit Error and Cell Loss Compensation Method for ATM Transport Systems*, Electronics and Communications in Japan, Part 1, Volume 76, Number 3
26. M.Bentall, B.Turton and C.Hobbs: *Improved Unassured Class C Services using a Modified Adaptation Layer 3/4*, Third Communication Networks Symposium, Manchester, 1996
27. M.Bentall: *Improved Network Resilience and its Applications to Asynchronous Networks*, PhD Thesis, University of Wales, Cardiff, 1997
28. ITU-T Q.931: *Digital Subscriber Signalling System No. 1 (DSS 1) - ISDN user–network interface layer 3 specification for basic call control*, March 1993
29. ITU-T Q.2100: *B-ISDN Signalling ATM Adaptation Layer (SAAL) Overview Description*, Helsinki, March 1993
30. ITU-T Q.2110: *B-ISDN ATM Adaptation Layer - Service Specified Connection Oriented Protocol (SSCOP)*, July 1994
31. ITU-T Q.2130: *B-ISDN Meta-Signalling Protocol*, February 1995
32. ISO-8348: *Information Technology – Open Systems Interconnection – Network Service Definition*, 1996
33. ITU-T X.213: *Information technology - Open Systems Interconnection - Network service definition*, November 1995
34. Richard Colella, Ella Gardner and Ross Callon: *Guidelines for OSI NSAP Allocation in the Internet*, RFC1237, July 1991
35. ISO-3166: *Codes for the Representation of Names of Countries and Their Subdivisions*, 1997
36. ITU-T Q.2120: Draft Recommendation, *B-ISDN Meta-signalling Protocol*, Paris, 1994
37. ITU-T Q.700: *Introduction to CCITT Signalling No.7*, Helsinki, March 1993
38. S.Aidarous and T.Plevyak (Editors): *Telecommunications Network Management into the 21st Century*, IEEE Press, 1993 (ISBN 0-7803-1013-6)
39. ITU-T G.707: *Network Node Interface for the Synchronous Digital Hierarchy (SDH)*, March 1996
40. ITU-T M.3010: *Principles for a Telecommunication Management Network*, Geneva, October 1992
41. ITU-T M.3020: *TMN interface specification methodology*, July 1995
42. ATM Forum: *Integrated Local Management Interface (ILMI) Specification Version 4.0*, af-ilmi-0065.000, September 1996
43. J.D.Case, M.Fedor, M.L.Schoffstall and C.Davin: *A Simple Network Management Protocol (SNMP)*, RFC1157, May 1990
44. S.G.Steinberg: *Netheads vs Bellheads*, Wired, October 1996
45. W.Almesberger, J.Le Boudec and P.Oechslin: *Application REQuester IP over ATM (AREQUIPA)*, RFC2170, July 1997
46. R.Braden, L.Zhang, S.Berson, S.Herzog and S.Jamin: *Resource ReSerVation Protocol (RSVP), Version 1 Functional Specification*, RFC2205, September 1997

7. L.Delgrossi and L.Berger: *Internet STream Protocol Version 2 (STII) Protocol Specification Version ST2+*, RFC1819, August 1995

8. W.Richard Stevens: *TCP/IP Illustrated, Volume 1*, Addison-Wesley Publishing Company, 1994

9. Christian Huitema: *The H Ratio for Address Assignment Efficiency*, RFC1700, October 1994

0. Yan-Jun Li and Stuart Elby: *TCP/IP Performance and Behaviour over an ATM Network*, NYNEX Science and Technology, White Plains, NY 10604, USA

1. Juha Heinanen: *Multiprotocol Encapsulation over ATM Adaptation Layer 5*, RFC1483, July 1993

2. M.Laubach: *Classical IP and ARP over ATM*, RFC1577, January 1994

3. ATM Forum: *LAN Emulation over ATM: Version 2 - LUNI Specification* (af-lane-0084.000), July 1997

4. ATM Forum: *ATM LAN Emulation Network–Network Interface (LNNI) Specification*, not yet issued

5. ATM Forum: *Multiprotocol Over ATM, Version 1.0* (AF-MPOA-0087.000), July 1997

6. J.V.Luciani, D.Katz, D.Piscitello and B.Cole: *NBMA Next Hop Resolution Protocol (NHRP)*, IETF draft-ietf-rolc-nhrp-11

7. A.Alles: *ATM Interworking*, Engineering Interop, Las Vegas, March 1995

8. P.Newman, G.Minshall and T.Lyon: *IP Switching: ATM Under IP*, Ipsilon Networks Inc., May 1997

9. P.Newman, W.Edwards, R.Hinden, E.Hoffman, F.Ching Liaw, T.Lyon and G.Minshall: *Ipsilon Flow Management Protocol Specification for IPv4, Version 1.0*, RFC1953, May 1996

0. K.Nagami, Y.Katsube, Y.Shobatake, A.Mogi, S.Matsuzawa, T.Jinmei and H.Esaki: *Toshiba's Flow Attribute Notification Protocol (FANP) Specification*, RFC2129, April 1997

1. P.Doolan, B.Davie, D.Katz, Y.Rekhter and E.Rosen: *Tag Distribution Protocol*, IETF draft-doolan-tdp-spec-01.txt, May 1997

2. A.Viswanathan, N.Feldman, R.Boivie and R.Woundy: *ARIS: Aggregate Route-Based IP Switching*, Draft RFC, September 1997

3. Y.Ohba, H.Esaki and Y.Katsube: *Comparison of Tag Switching and Cell Switch Router*, IETF draft-ohba-tagsw-vs-csr-00.txt, April 1997

4. Y.Rekhter, B.Davie, D.Katz, E.Rosen and G.Swallow: *Cisco Systems' Tag Switching Architecture Overview*, RFC2105, February 1997

5. P.Newman, W.Edwards, R.Hinden, E.Hoffman, F.Ching Liaw, T.Lyon and G.Minshall: *Ipsilon's General Switch Management Protocol Specification*, RFC1987, August 1996

6. Y.Katsube, K.Nagami and H.Esaki: *Toshiba's Router Architecture Extensions for ATM: Overview*, RFC2098, February 1997

7. ITU-T Q.2931: *Broadband Integrated Services Digital Network (B-ISDN) - Digital subscriber signalling system no. 2 (DSS 2) - User–network interface (UNI) - Layer 3 specification for basic call/connection control*, February 1995

8. Motorola: *The Basics Book of Frame Relay*, Motorola University Press, 1995 (ISBN 0-201-56377-0)

9. U.Black: *Frame Relay Networks: Specifications and Implementations*, McGraw-Hill, 1996 (ISBN 0-07-005590-4)

0. Frame Relay Forum: *Frame Relay/ATM PVC Network Interworking Implementation Agreement*, December 20th 1994, FRF.5

71. Frame Relay Forum: *Frame Relay/ATM PVC Service Interworking Implementation Agreement*, December 20th 1994, FRF.8
72. ITU-T Q.922: *ISDN Data Link Layer Specification for Frame Mode Bearer Services*, February 1992
73. P.Wouters: *Designing a Safe Network Using Firewalls*, LINUX Journal, August 1997
74. J.Moy: *OSPF Version 2*, RFC2178, July 1997
75. A.Tanenbaum: *Network Protocols*, Computing Surveys, Volume 13, Number 4, December 1981
76. T.Berners-Lee, L.Masinter and M.McCahill: *Uniform Resource Locators (URL)*, RFC1738, December 1994

Index